ADVANCES IN RENEWABLE ENERGY RESEARCH

Advances in Renewable Energy Research

Małgorzata Pawłowska & Artur Pawłowski
Lublin University of Technology, Lublin, Poland

CRC Press is an imprint of the
Taylor & Francis Group, an **informa** business
A BALKEMA BOOK

CRC Press/Balkema is an imprint of the Taylor & Francis Group, an informa business

Typeset by V Publishing Solutions Pvt Ltd., Chennai, India
Printed and bound in Great Britain by Antony Rowe (A CPI-group Company), Chippenham, Wiltshire

Published by: CRC Press/Balkema
Schipholweg 107C, 2316 XC Leiden, The Netherlands
e-mail: Pub.NL@taylorandfrancis.com
www.crcpress.com – www.taylorandfrancis.com

ISBN: 978-1-138-55367-5 (Hbk)
ISBN: 978-1-315-14867-0 (eBook)

Table of contents

Foreword

Without the supply of an adequate amount of energy, the present civilization could not exist. The technological revolution, which initiated the on-going development of the world, was and is based mainly on the use of fossil fuels as a source of primary energy.

There is a growing concern that increasing levels of greenhouse gases in the atmosphere, particularly CO_2, are contributing to global climate change (IPCC, 2007). Atmospheric levels of CO_2 have risen significantly from preindustrial levels of 280 parts per million (ppm) to present levels of 384 ppm. Evidence suggests that elevated atmospheric CO_2 concentrations are the result of a combination involving expanded use of fossil fuels for energy production and transportation, land use conversion (deforestation), and soil cultivation.

The threats caused by climate change resulting from the emission of greenhouse gases, especially carbon dioxide, are well-known. Less known, however, are the threats connected with rapidly depleting fossil fuels. According to La Quere (La Quere et al. 2015), the fossil fuel resources are being quickly depleted. Forecasts predict that the available deposits of gas and crude oil will be exhausted within the next 50–70 years, whereas the deposits of coal – will be depleted within 130–150 years. This means that there is an urgent need of acquiring new sources of primary energy.

Therefore, developing renewable energy sources is necessary not only from the viewpoint of climate protection, but also to ensure that new energy sources are available for the functioning of civilizations. Within the period of 1950–2016, the consumption of primary energy increased from 36.52 exajoules in 1950 to 102.73 exajoules in 2016. The consumption of energy from renewable sources increased from 3.14 exajoules in 1950 to 10.82 exajoules in 2016. This means that with the total increase of the primary energy in 1950–2016 by 281%, the consumption of energy from renewable sources increased with 341%, including the consumption of energy derived from biomass which increased from 1.65 exajoules in 1950 to 5.02 exajoules in 2016, i.e. 305%. On the other hand, the share of biomass in the pool of renewable energy, including biogas, dropped from 52% in 1950 to 46.8% in 2016. Nevertheless, the energy acquired from biomass still constitutes almost half of the energy derived from renewable sources.

The largest producer of biofuels is the USA, which produces approximately 35,000 metric tons of oil equivalent per year. The second most important producer of biofuels is Brazil, with an output of 18,500 metric tons of oil equivalent per year. In EU countries, Germany is the largest producer of biofuels, yielding 3,200 metric tons of oil equivalent per year, followed by France, producing 2,200 metric tons of oil equivalent per year. A great interest in the production of biofuels is observed in China, which is the 7th country in the world in terms of biofuel production, yielding slightly more than 2,000 metric tons of oil equivalent per year.

The use of biomass as a source of primary energy differs greatly in developing and developed countries. In the former, biomass – mainly wood – is one of the basic sources of primary energy. On the other hand, in the developed countries, advanced methods of biomass use for energy purposes are being developed, especially for the production of biogas and liquid biofuels: ethanol and biodiesel. Biomass from special plantations is usually used for the production of these advanced biofuels.

However, many uncertainties remain for the future of biofuels, including competition from unconventional fossil fuel alternatives and concerns about environmental tradeoffs. Perhaps the biggest uncertainty is the extent to which the land intensity of current biofuel production can be reduced.

The development of biofuels, while reducing the demand for fossil fuels, may also constitute a threat. Allocating increasing areas of land for the cultivation of subsidized energy crops leads to the limitation of food production, and consequently, to an increase of food prices which can already be noticed. Moreover, negative environmental consequences have been observed, e.g. elimination of tropical forests by the energy crops plantations. This may even lead to an increased CO_2 emission, because the tropical forests are replaced with energy crops, which absorb a smaller amount of CO_2.

This does not mean that the development of currently used biofuels is a negative phenomenon. However, conducting a full LCA of each technology to be used is necessary in order to eliminate the ones which do not achieve positive results in terms of environmental protection.

During the conference, researchers from Germany, Poland, Ukraine, China, Japan and Taiwan presented the results of works carried out in these countries related to the production of biogas, bioethanol and biodiesel from wastes and purposely cultivated crops.

REFERENCES

IPCC, 2007: *Climate Change 2007: Synthesis Report. Contribution of Working Groups I, II and III to the Fourth Assessment Report of the Intergovernmental Panel on Climate Change* [Core Writing Team, Pachauri, R.K. and Reisinger, A. (eds.)]. IPCC, Geneva, Switzerland, 104 pp.

La Quere, C. et al. 2015, Global Carbon Budget 2015, Earth Syst. Sci. Data, 7, 349–396, 2015.

International Center for Renewable Energy (ICERN) Memorandum of Understanding (MoU)

MISSION

The International Center for Renewable Energy (ICERN) is an independent non-governmental organization based in Lublin, Poland. The ICERN is a center of excellence in renewable energy that supports and promotes development and application of technologies to produce energy from renewable sources and solid and liquid wastes.

The ICERN mission is also to broaden the base of expert input on the development of effective public policy for renewable energy and to provide a channel through which leading scientists in the field of renewable energy community can have their views known.

ICERN MAJOR TOPIC AREAS

The following areas of special interest for ICERN are:

1. Development of bio-based technologies to produce energy from solid and liquid wastes.
2. Development of methods for biological sequestration of CO_2
3. Development of solar energy technologies for electricity and cooling/heating
4. Perspective technologies of wind energy
5. Development of geothermal energy technologies
6. Technology of small hydro power
7. Development and application of method for renewable energy technologies, including LCA
8. Development of pretreatment methods of biomass to enhance biogas production
9. To setup a subcenter of ICERN in China for PV-agriculture Research and Development in NAU
10. Development of design and planning ability for coordination and balance between renewable energy industry and other industries
11. Interaction of remediation technology and fertilizer cycles in bioenergy systems

Promotion of research and development within the above-mentioned areas has technology and industry scope. It focuses on:

- thought leadership,
- positive positioning of renewable energy bringing it into the mainstream of energy production in the EU and other countries,
- benefits to the nations in the context of environment and human welfare

ICERN ACTIVITIES

The ICERN program is implemented through:

- joint partner application for research and industrial funds,
- development of educational modules,
- organization of promotion events, such as specialty symposia, conferences

- exchange of faculty, postdoc and students
- organization of Network of Informational Centres of energy efficiency an technology transfer of renewable energy

ICERN MANAGEMENT

The ICERN founding members include:

- **Prof. Zhexenbek M. Adilov**, Kazakh National Research Technical University, Almaty
- **Prof. January Bień**, Częstochowa Univeristy of Technology, Częstochowa
- **Katarzyna Bułkowska**, University of Warmia and Mazury, Olsztyn
- **Prof. Zhihong Cao**, Institute of Soil Sciences, Chinese Academy of Sciences, Nanjing
- **Yuriy Favorskyy**, Institute of Renewable Energy National Academy of Sciences of Ukraine, Department of Renewable Energy at the National Technical University of Ukraine "KPI", Kiev
- **Mariusz Gusiatin**, University of Warmia and Mazury, Olsztyn
- **Prof. Jan Kiciński**, Institute of Fluid-Flow Machinery, Polish Academy of Science, Gdansk
- **Prof. Aldebergen Malibekov**, M.Kh. Dulaty Taraz State University, Taraz
- **Prof. Agnieszka Montusiewicz**, Lublin University of Technology, Lublin
- **Ph.D. Eng. Elisabeth Pacyna**, NILU – Norwegian Institute for Air Research, Kjeller
- **Prof. Józef Pacyna**, NILU – Norwegian Institute for Air Research, Kjeller
- **Prof. Małgorzata Pawłowska**, Lublin University of Technology, Lublin
- **Prof. Artur Pawłowski**, Lublin University of Technology, Lublin
- **Prof. Lucjan Pawłowski**, Lublin University of Technology, Lublin
- **Oleksandr Pepelov**, SE Inter-branch Science and Technology Center for Windpower at NAS of Ukraine, Institute for Renewable Energy of NAS of Ukraine, Kiev
- **Prof. Mattnetkali Sarybekov**, M.Kh. Dulaty Taraz State University, Taraz
- **Prof. Heralt Schöne**, Hochschule Neubrandenburg, Neubrandenburg
- **Prof. Witold Stępniewski**, Lublin University of Technology, Lublin
- **Prof. Jung-Jeng Su**, National Taiwan University, Taipei
- **Prof. Junichi Takahashi**, Obihiro University of Agriculture and Veterinary Medicine, Obihiro, Hokkaido
- **Prof. Waldemar Wójcik**, Lublin University of Technology, Lublin
- **Prof. Yanwen Zhao**, Nanjing Agricultural University, Nanjing
- **Prof. Mirosław Żukowski**, Bialystok University of Technology, Bialystok

The ICERN founding members form a Management Board of ICERN that is signatories of this MoU.

The secretary of the ICERN activities is located at the Lublin University of Technology.

About the editors

Małgorzata Pawłowska, Ph.D., Sc.D. (habilitation) was born in 1969 in Sanok, Poland. In 1993 she received M.Sc of the protection of the environment at the Catholic University of Lublin. Since that time she has been working in the Lublin University of Technology, Faculty of Environmental Engineering. In 1999 she defended her Ph.D. in the Institute of Agrophysics of the Polish Academy of Science and in 2010 she defended her D.Sc. thesis at the Technical University of Wroclaw and was appointed as associate professor and head of Engineering of Alternative Fuels Department at the Faculty of Environmental Engineering, Lublin University of Technology. Now she is working on biomethanization processes and the application of selected wastes for remediation of degraded land.

She has published 60 papers, 3 books and is co-author of 12 Polish patents and 4 European patents.

Artur Pawłowski, Full Professor, was born in Lublin in 1969. From 1993 he has been working in the Lublin University of Technology. He is Head of the Department of Sustainable Development. He works on issues related to environmental engineering, renewable sources of energy and multidimensional nature of sustainable development. In 2011 he published the book "Sustainable Development as a Civilizational Revolution. A Multidisciplinary Approach to the Challenges of the 21st Century" (CRC Press). He is a member of European Academy of Science and Arts, Salzburg, Austria, The Committee of Environmental Engineering of the Polish Academy of Sciences, Warsaw, Poland, International Academy of Ecological Safety and Nature Management, Moscow, Russia and International Association for Environmental Philosophy, Philadelphia, United States. He is editor-in-chief of the scientific journal: Problemy Ekorozwoju/ Problems of Sustainable Development and member of the editorial board of the Committee of Environmental Engineering monographs.

He is the author of more than 150 publications, published in English, Polish and Chinese.

About the authors

Fast methanification of liquid pig manure as an example for substrates with low organic content*

H. Schöne & I.B. Mendes de Oliveira
Department of Agriculture and Food Science, Hochschule Neubrandenburg, Neubrandenburg, Germany

A. Speetzen
ME-LE-Energietechnik GmbH, Torgelow, Germany

ABSTRACT: A biogas reactor of 45 m^3 was fed with pure liquid pig manure. A straw layer served as an anaerobic filter on top of the fluid. The manure was continuously circulated to irrigate the straw. Hydraulic Retention Time (HRT) of straw was 45 days. HRT of manure was reduced from 45 to 7.5 days within one year. Average concentration of Volatile Solids (VS) of manure was only 1.8%.

We varied VS concentration and temperature to simulate normal disturbances of operation. Gas production normalized within one day after each short heating interruption. Variations of VS concentration had no negative influence on the operation as a whole.

After two months, a zone with granular sludge in autonomous fluidization was observed just below the straw layer. This shows that the reactor is a hybrid biogas reactor containing a fixed bed on the top, and an UASB zone below.

Keywords: Anaerobic digestion; biogas; swine waste; hybrid reactor; UASB

1 INTRODUCTION

One difficulty of biogas production from swine manure is that the concentration of volatile solids (VS) frequently falls between 1% and 2% (m/m). We measured a yield of 205 l methane per kg VS from swine manure over 11 months. If calculated with a net heating value of 36 MJ per m^3 methane and 1.8% VS (m/m), we have 18 kg VS per m^3 manure yielding 3.69 m^3 methane and 132.84 MJ. In order to heat 1 m^3 manure from 20°C to 35°C, a heat energy $Q = 1000 \text{ kg} \times 4.2 \text{ kJ/(kg K)} \times 15 \text{ K} = 63 \text{ MJ}$ is required (the manure can be treated as water in this calculation). That is approximately equivalent to all the heat that can be supplied from a combined heat and power plant (CHPP) fed by biogas (approx. 40% of the calorific energy of the methane when 40% is electric energy, 40% heat and 20% losses).

This means that a biogas plant running with pure manure can supply its own heat requirement under normal conditions, but in cold winters there may be a shortage. Ideally, the feed is passed through a heat exchanger with the effluent in countercurrent flow and the digester itself is well insulated. This is easy to perform if the digester is small.

For the digester volume to be small, the hydraulic retention time (HRT) of the manure needs to be short. The doubling time of methane-forming archaea is in the range between 10 and 12 days. That means HRT should be at least 12 days in order not to wash out the microorganisms. In contrast to this finding, Hill and Bolte (Hill and Bolte, 2000) published the results showing that a digester of the continuously stirred reactor type (CSTR) can be run steadily with a HRT of three days. This result can be explained because the fresh manure has

---

* First published in: Schöne, H.; Speetzen, A.; Bernardes, I.; de Oliveira, M. (2015): Fast methanification of swine manure as an example for substrates with low organic content, in: Nelles, M (Hrsg.) Tagungsband zum 9. Rostocker Bioenergieforum 2015, 18.–19.06.2015, S. 303–308, ISBN 978-3-86009-425-9.

already been hydrolyzed, and microorganisms are continuously taken into the digester with the manure. In our estimation, this cannot be achieved. Therefore, there should be a reactor type which prevents microorganisms from being washed out. To achieve stable operation under changing conditions (e.g. temperature, VS content) the alternative was to construct a fixed bed reactor with a fixed bed consisting of fibrous material.

2 METHODS

We constructed a cylindrical digester of the size height × diameter = 5.5 × 3.4 m with a volume V = 50 m^3. A headspace of 5 m^3 is for collecting biogas, so the operational volume is 45 m^3 (Figure 1). A more detailed description is given in EP 2 314 666 A1 (Rossow, 2009). The digester was run with pure swine manure from a stable located nearby, under mesophilic conditions. The stable houses 1200 pigs. This pilot plant is located on a swine farm site close to Neubrandenburg, and is fully integrated into the infrastructure of a biogas plant of commercial dimensions. That means that the heating is part of the warm water circuit supplied from the CHPP nearby, and the biogas is used in this plant. The feed of our test digester is supplied from the existing tubing system which feeds the commercial plant.

A wheat straw layer which has been chopped to a length of five centimeters floats on top of the fluid in the testing digester, acting as an anaerobic filter. The layer material was prepared according to findings of Wilkie and Colleran (Wilkie and Colleran, 1984) which show that swine manure can be treated with anaerobic filters, and Andersson and Björnsson (Anderson and Björnsson, 2002) demonstrating that straw is an excellent biofilm carrier for the digestion of crop residues compared to a glass packed bed, and a reactor containing suspended plastics. Our principle of operation requires the straw to float upon the fluid.

To find out the best combination of straw and chop length, we processed wheat straw, rye straw, and barley straw in preliminary batch digestion experiments in 0.5l-bottles at 38°C with swine manure. Of these types of straw, the wheat straw exhibited the best floating behavior.

In the start-up phase, the 45 m^3-reactor was filled with swine manure and heated to 38°C. Then, wheat straw (chopped) was added continuously to build up the straw layer. The straw was fed into a feeder which was integrated in the recirculation tubing.

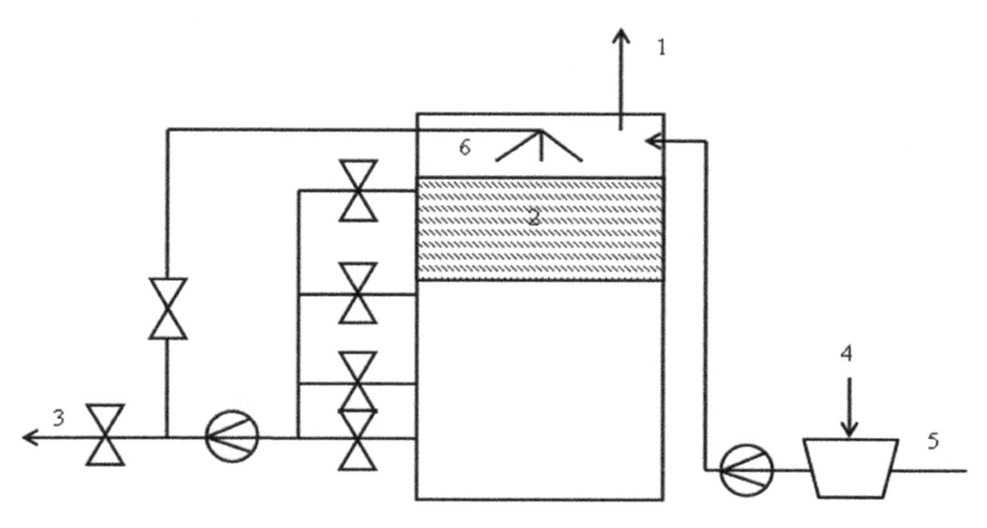

Figure 1. Scheme of pilot-scale biogas reactor described in the text. 1 – Biogas, 2 – straw layer, 3 – effluent, 4 – straw feed, 5 – manure feed, 6 – sprinkler system.

After the start-up phase, the swine manure feed mass flow was gradually increased, which steadily reduced the hydraulic retention time of the manure. Furthermore, the daily amount of straw was gradually reduced until equilibrium was reached (consistent and stable floating layer).

The manure was circulated in the reactor so the straw layer was sprinkled continuously. The hydraulic retention time (HRT) of the straw was approx. 45 days. The HRT of the manure started with 45 days and was reduced to 7.5 days over a period of 11 months. The average concentration of volatile solids (VS) of the swine manure was 1.8 mass percent and changed during the experiment, as shown in Figure 2; this figure shows VC concentration and Figure 3 the HRT over time. The characteristics of the manure are summarized in Table 1.

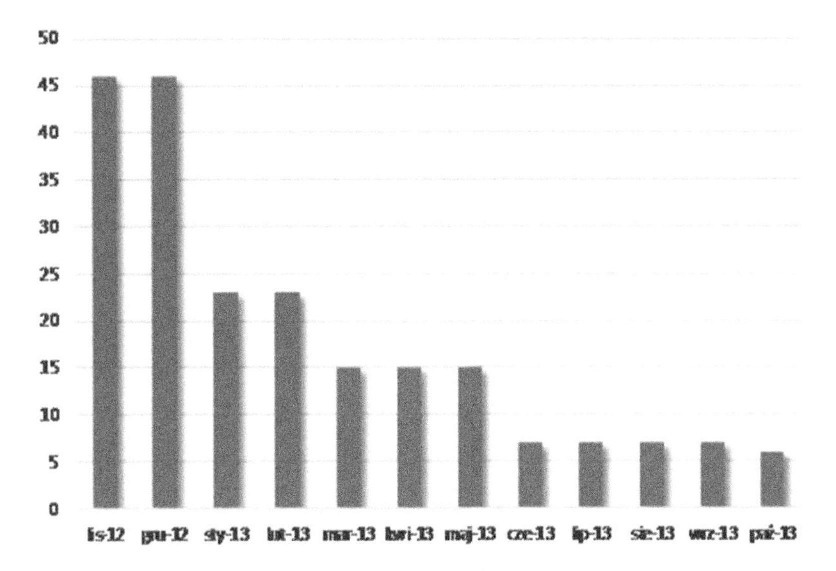

Figure 2. Volatile Solids [m/m%] over time, broken line: average.

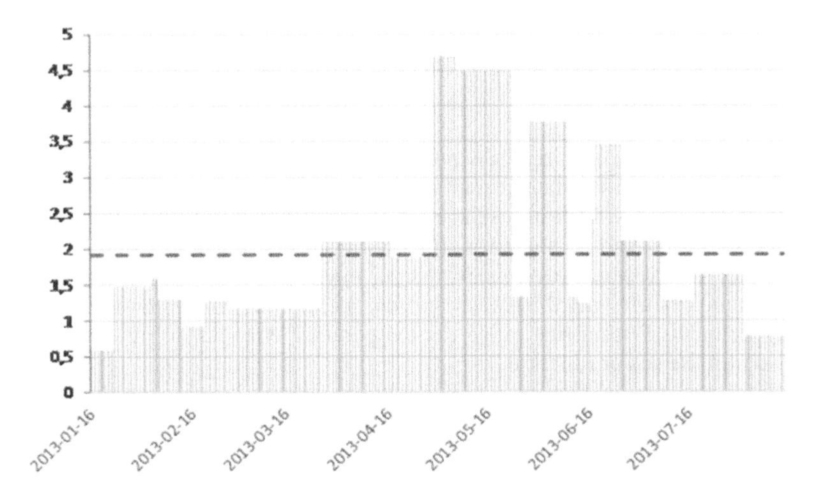

Figure 3. Hydraulic Retention Time (HRT) [d] over time.

Table 1. Substrate—General Parameters: minimum, maximum and mean values.

Parameter	Minimum	Mean	Maximum
pH	7.4	7.7 ± 0.2	8.1
FOS/TAC (Organic Volatile Acids/Total Inorganic Carbonate)	0.12	0.3 ± 0.1	0.55
Total Solids (Dry Matter) (% m/m)	1.2	3 ± 1	6.31
Volatile Solids (Organic Dry Matter) (%) of TS	56	65 ± 5	73.9
Volatile Solids (Organic Dry Matter) (% m/m)	0.67	1.6 ± 0.7	3.78
Fixed Solids (%)	26.1	35 ± 5	44
Fixed Solids (% m/m)	0.46	0.8 ± 0.2	1.34
Concentration of acetic acid (mg/L)	92.5	3000 ± 967	1397
Concentration of propionic acid (mg/L)	50	267 ± 209	1411
Propionic Acid/Acetic Acid	0.07	0.2 ± 0.1	0.54

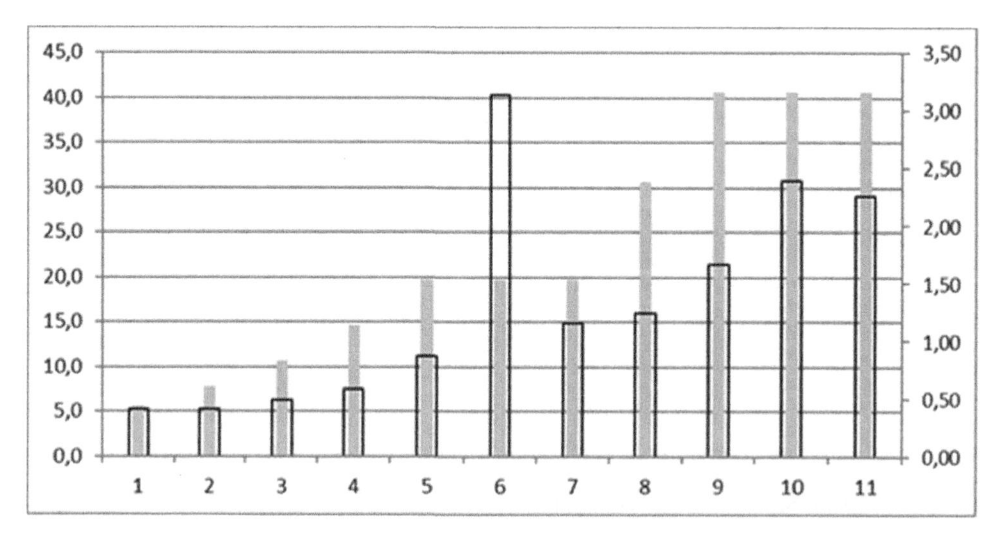

Figure 4. Biogas production [m³/day] (bold line, left scale), temperature [°C] (shadow, right scale) over time.

3 RESULTS AND DISCUSSION

Reactor behavior was stable. During the experiment we varied the VS concentration and the temperature several times to simulate normal operational disturbances. After each short heating interruption, the gas production rose again to normal values within one day. The variation of VS concentration had no negative influence on the operation as a whole.

Figure 4 shows the gas production measured per day and the temperature in the digester over time. The temperature varied between approx. 31.5°C (minimum) and approx. 42.5°C during the experiment. Within this range the temperature changes do not seem to have a negative impact on the gas production.

Figure 5 shows the organic loading rate (OLR) over time and the gas production per day over time. This figure demonstrates that the gas production rose with the OLR over time, and there was no OLR which was high enough to negatively influence productivity.

During the experiment, there was no clogging of the straw layer. Despite a decreasing HRT, the biogas yield raised from 300 (45 days HRT) to more than 500 L/kg VS (11–9 days HRT).

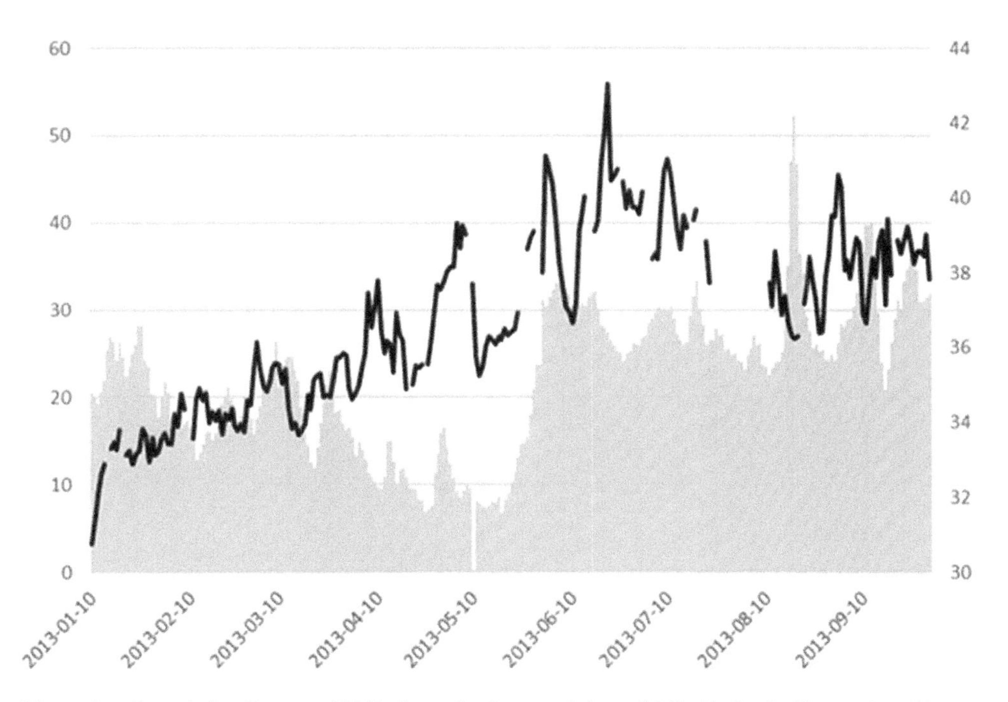

Figure 5. Organic loading rate (OLR, framed columns, right scale) [kg/(m³ × day)] over time, biogas production [m³/day] over time (grey columns, left scale).

4 CONCLUSIONS

This reactor type can be used for the processing of swine manure on a large scale with low investment costs. The solution is suitable for pig farms all over the world, especially in emerging countries. This type of a reactor is very stable in terms of short-term changes in VS concentration and temperature. We estimate that it is suitable for other substrates with low solids content as well.

ACKNOWLEDGEMENTS

We thank the state of Mecklenburg-Western Pomerania for funding this research with EFRE and ESF subsidies from the European Union.

REFERENCES

Andersson, J. & Björnsson, L. 2002. Evaluation of straw as a biofilm carrier in the methanogenic stage of two-stage anaerobic digestion of crop residues. *Bioresource Technology*, 85(1): 51–56.

Hill, D.T. & Bolte, J.P. 2000. Methane production from low solid concentration liquid swine waste using conventional anaerobic fermentation. *Bioresource Technology*, 74(3): 241–247.

Rossow, N., EP 2314666 A1, patent application, Method and device for increasing the population density of bacterial cultures which produce methane in biogas fermenters, priority date Oct. 21st, 2009.

Wilkie, A. & Colleran, E. 1984. Start-up of anaerobic filters containing different support materials using pig slurry supernatant. *Biotechnology letters*, 6(11): 735–740.

Development and demonstration of farm scale biogas biofilter systems for livestock biogas applications in Taiwan

J.J. Su
Faculty of Animal Science and Technology, Division Chief of Bioenergy Research Center, National Taiwan University, Taipei, Taiwan

ABSTRACT: The farm scale biogas bio-filter system (BBS) is serially developed by enrichment and isolation of sulfide-oxidizing bacteria, pilot scale bio-filter reactor establishment and testing, as well as farm scale bio-filter facility establishment and testing for more than 3 yrs. Sulfide dioxide emission is also monitored in three commercial pig farms which utilize two types of biogas, either bio-desulfurized or un-desulfurized. Demonstration of farm scale BBS facilities is carried out on a 25000-head commercial pig farm which was selected as an official demonstration site for biogas power generation in 2015 by Bureau of Energy, Ministry of Economic Affairs, Taiwan. Furthermore, this farm-scale BBS with three patents was conferred an award at the 7th National Innovation Award, Taiwan in 2010.

Keywords: livestock biogas; biofilter; bio-desulfurization; renewable energy

1 INTRODUCTION

Biogas produced from anaerobic piggery wastewater treatment plants in Taiwan contains methane (60.1–77.0%), carbon dioxide, a low proportion of nitrous oxide, and hydrogen sulfide (Su et al. 2003). Global warming potential (GWP) indicates the heat-trapping potential of greenhouse gases in the atmosphere. The GWP of methane (CH_4) is 25, i.e. 25 times more potent than carbon dioxide (CO2) in causing global warming (IPCC 2007). Thus, promotion of biogas utilization or combustion can help to reduce greenhouse gas emissions (especially CH4) from the livestock sector (Velkin and Shcheklein 2017, Cao et al. 2016, Cel et al. 2016). Biogas produced from livestock farming is an alternative energy source in several countries (Omer & Fadalla 2003, Kapdi et al. 2005, Murphy & McCarthy 2005, Prasertsan & Sajjakulnukit 2006, Craggs et al. 2008, Ihara et al. 2008, Lieffering et al. 2008), but power generators normally demand that the utilized biogas contain less than 280 mg/m^3 of H2S. The anaerobic wastewater treatment process used in the paper industry can produce as high as 42000 mg/m^3 H_2S (Schieder et al. 2003), while that of biogas from anaerobic piggery wastewater treatment is up to 5600 mg/m^3 in Taiwan. Therefore, the high content of H_2S limits the use of biogas for heat and power generation.

Hydrogen sulfide can be removed from biogas either by non-microbial or microbial processes. Non-microbial processes include either dry oxidation or liquid phase oxidation processes (Kapdi et al. 2005), while microbial processes use certain photoautotrophic and chemotrophic bacteria immobilized in bioreactors to remove H_2S from gas streams (Kantachote et al. 2008).

Certain photoautotrophic and chemotrophic bacteria immobilized in bioreactors remove H_2S from gas streams (H_2S S^0 SO_4^{2-}) (Kantachote et al. 2008). The five most effective microbiological devices developed to remove H_2S from gas streams are gas-fed batch reactor, continuous stir-tank reactor (CSTR), bio-scrubber, bio-filter, and a bio-trickling filter (Syed et al. 2006). Both gas-fed batch reactor and CSTR for H_2S removal require electricity for oxygen supply, continuous stirred liquid, and lighting (Kim et al. 1996; Basu et al. 1996; Janssen et al. 1995).

Both bio-scrubber system (fixed film) and bio-trickling system for H2S removal still require electricity for recycling of liquid and replacing it with fresh one (Potivichayanon et al. 2005, Moosavi et al. 2005). Hydrogen sulfide (20–6500 mg/m^3) removal efficiencies are reported to achieve 85–99% using either a pilot- or bench-scale bio-trickling filter (Moosavi et al. 2005). Pilot or bench-scale bio-filters are used to treat H_2S (3.2–1300 mg/m^3) in biogas and achieve H_2S removal efficiency of 29.7–100% with various bio-carriers (Moosavi et al. 2005). Notably, no bio-filter in these studies was farm-scale.

Two stages of H_2S removal in systems using bio-scrubbers (fixed film) are absorption of H_2S by a liquid and biological oxidation of H_2S in the liquid (Potivichayanon et al. 2005). The bio-filter is a three phase bioreactor (gas, liquid, and solid) with a filter bed that has a high porosity, buffer capacity, nutrient availability, and moisture retention capacity to ensure that the target bacteria can grow on it. The novelty of this work involves developing and demonstrating a high efficiency and cost-effective farm-scale biogas bio-desulfurization system (BBS) without recirculating and aerating flushing liquid.

From 2009 to 2014, farm-scale BBS had been established in certain commercial pig farms with the pig population up to 25000 heads. Out of these farms, a 25000-head pig farm has been selected as a demonstration site for biogas power generation in 2015 by Bureau of Energy, Ministry of Economic Affairs, Taiwan.

2 ISOLATION OF SULFIDE OXIDIZING MICROORGANISMS

A pilot-scale biogas biofilter reactor with enrichment cultures was designed and operated on a 9000-head commercial pig farm in Miaoli, Taiwan for more than a year. The hydrogen sulfide removal efficiency of the pilot-scale biogas bio-filter system (BBS) (biogas flow rate = 4 L/min) in the pig farm was up to 99%. Sulfur oxidizers can convert sulfide (S2–) to sulfur (S0) and even sulfate (SO_4^{2-}). CYAS-1 and CYAS-2 strains, both characterized by a significant sulfide oxidation capability, were isolated from the sludge of piggery wastewater treatment facilities. Moreover, SW-1, SW-2, and SW-3 strains were isolated from a pilot-scale biogas bio-filter reactor. All isolates were then proven to have capabilities of sulfide oxidation by growing them in 150 mL liquid media with 1.5 g sulfur powder. An increase in sulfate concentration was used to select sulfide oxidizers. Experimental results showed that CYAS-1strain (identified as Acinetobacter spp.), grown in diluted trypticase soy broth (TSB) with sulfur powder, increased the concentrations of SO_4^{2-} from 17.2 ± 0.5 to 23.8 ± 1.0 mg/L (38.4% increase).

CYAS-2 Strain (identified as Corynebacterium spp.), grown in diluted TSB with sulfur powder, increased the concentrations of SO_4^{2-} from 17.7 ± 0.1 to 25.9 ± 0.9 mg/L (47.0% increase). The concentrations of SO_4^{2-} were increased 40.5, 33.6, and 29.7% in the presence of SW-1 (Candida kruse/inconspicua; 96.2% identity), SW-2 (Candida parapsilosis; 93.2% identity), and SW-3 (Trichosporon mucoides; 95.7% identity) strains, respectively (Su et al. 2008).

3 DEVELOPMENT OF PILOT-SCALE BIO-DESULFURIZATION SYSTEM

This study analyzed the performance of a pilot-scale biogas bio-filter system (BBS) operating for more than 630 days on a 9000-pig commercial farm equipped with anaerobic digesters and biogas storage bags in Miaoli, Taiwan (Su et al. 2014). The main objective was to evaluate the efficiency of the BBS facility in removing H_2S and to identify the optimal operating parameters for efficient removal of H_2S from biogas.

3.1 *Pilot-scale biogas bio-filter system (BBS)*

The pilot-scale BBS was wrapped with thermal insulation layers and installed beside two plug flow anaerobic digesters of the same commercial pig farm. The BBS consisted of two separate acrylic plastic columns (internal diameter (i.d.) 0.2 × 1.0 m) in series (Figure 1a). Each 62.8-L column was packed with dried aerial roots (50.2 L), and flow rates of 4 L/min and 6 L/min were tested in order to find the optimal flow rate for H_2S removal from biogas.

A 0.5HP air vacuum pump (TECO Electric & Machinery Co., Taiwan) was installed at the outlet of the BBS to withdraw biogas from the storage bags and introduce it into the BBS by negative pressure. The biogas produced from the commercial pig farm anaerobic digester was introduced to the inlet of the BBS using PVC pipelines. In this study, oxygen, essential for growing sulfur-oxidizing bacteria, was introduced into the BBS through an annular gap in the connector of the biogas inlet pipeline by removing an O-ring from the connector. From the stoichiometry of sulfide oxidation, the oxygen supplied enabled the oxidation of both sulphide ($4H_2S + CO_2 + O_2 \rightarrow CH_2O + 4S + H_2O$) and sulfur ($2S + 3O_2 + 2H_2O \rightarrow 2H_2SO_4$).

A humidifier (i.d. 200 × 450 mm) was also installed to ensure that the humidity of biogas entering the BBS was 85% or higher, followed by the biogas bio-filter unit. The humidifier had a working volume of 14 L and was filled with 8 L of clean tap water.

3.2 *Inoculation and time-course experiments of pilot-scale BBS*

Two bio-carriers used in the current study were dried aerial roots of the common tree fern (Sphaeropteris lepifera) and small plastic rings. The dried aerial roots were about 30–70 mm long and are commonly used as carriers for cultivating orchids in Taiwan. The small plastic Rasching rings (double spherical shape; 32 × 38 mm) (Sheng-Fa Plastics Inc., Touyuan, Taiwan) were made of polypropylene (PP) plastic. The roots were soaked in a mixture of activated sludge from pig farms, pond water and soils. After soaking, the aerial roots were screened through a rough-mesh (5 mm) sieve and packed into the reactor.

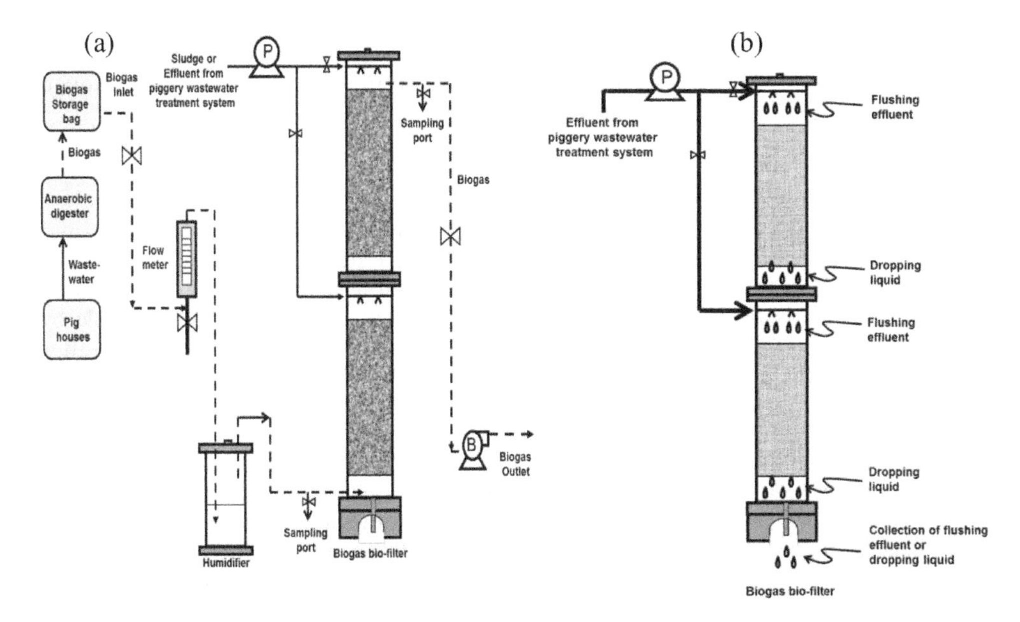

Figure 1. a) Sketch of the bio-filter setup. b) Sketch of the sampling sites of liquid samples from the BBS (Su et al. 2014).

Table 1. Description of liquid samples for ion chromatography analysis (Su et al. 2014).

Liquid samples	Description
Effluent	Treated wastewater discharged from a wastewater treatment system
Flushing effluent	The effluent after it is flushed out of the biofilter
Dropping liquid	Condensed liquid from bio-carriers inside the biofilter
Soaking liquid	The supernatant of biofilm-coated aerial roots soaking in deionized water

Table 2. Sulfate, ammonium, and pH (± S.D.) in three liquid samples from BBS by ion chromatography (Su et al. 2014).

Samples	$NH4^+$ (mg/l)		$SO4^{2-}$ (mg/l)		pH	
Effluent	112	8.7	65	13.3	8.0	0.1
Dropping liquid	112	24.9	28 680	8253	1.6	1.0
Flushing effluent	77	16.3	9054	1043	2.4	1.4
P	NS		0.01		0.01	

Data presented as mean S.D.
NS: Not significant.

For the later time-course experiment, a mixture of dried aerial roots (35.3 L) and small plastic rings (15 L) was used as the bio-carrier. Biogas was introduced into the bio-filter system through a flow meter (Tohama 10B, Yeong Shin Co., Taiwan), a humidifier, and the BBS before being withdrawn with a 0.5HP air vacuum pump. The oxygen content was about 7% (v/v), measured at the outlet by means of a portable multi-gas leak detector (MX6 iBRID, Industrial Scientific Co., USA) with a resolution of 0.1%. All microorganisms used for the current study were isolated and identified for the capability of sulfur oxidation, based on experimental results (Su et al. 2008). The BBS was operated for 120 days before performing a 100-day continuous time-course experiment to evaluate the potential of the different packing materials in removing H2S (Su et al. 2013). Samples (12 mL) of the liquid from bio-filter operation and the soaking liquid (Table 1) were collected (Figure 1b) and analyzed through ion chromatography (IC) (Su et al. 2008).

3.3 *Removal efficiency of hydrogen sulfide*

For Days 1–99, the relative humidity of inlet biogas was in the range of 79.3–99.8% and the temperature amounted to 15.2–38.4°C. Outlet relative humidity reached 75.5–99.8%, and temperature was 23.1–38.0°C. Inlet H_2S concentrations were 1540–5600 mg/m^3 and in some instances possibly even higher, since the maximum concentration detectable by H_2S tubes is only 5600 mg/m^3. Concentrations of H_2S in the biogas at the outlet, after passing through the bio-filter, were in the range of 0–2240 mg/m^3 and removal efficiency of H_2S amounted to 27.3–100% (removal efficiency of H_2S was calculated based on inlet H_2S = 5600 mg/m^3).

During Days 62–97, the BBS flow was fixed at 4 litre/min, but the elemental sulfur (S^0) still accumulated between and on the surfaces of bio-film-coated aerial roots (i.e. H_2S S^0). The excess elemental sulfur destabilized the biogas de-sulfurization by the BBS. To minimize elemental sulfur accumulation on aerial root surfaces, a different bio-carrier, small Rasching rings (32 × 28 mm), was used as a second bio-carrier for the bio-filter on Day 98. On Days 99–217, the average inlet relative humidity of biogas was 99.9% and the temperature range was 20.8–39.0°C. The outlet relative humidity reached 52.8–92.7%, and the temperature amounted to 20.6–34.3°C. The experimental results showed that the efficiency of H_2S removal was 90.0–100% (Figure 2). The H_2S concentrations in the inlet biogas were 2800–5600 mg/m^3

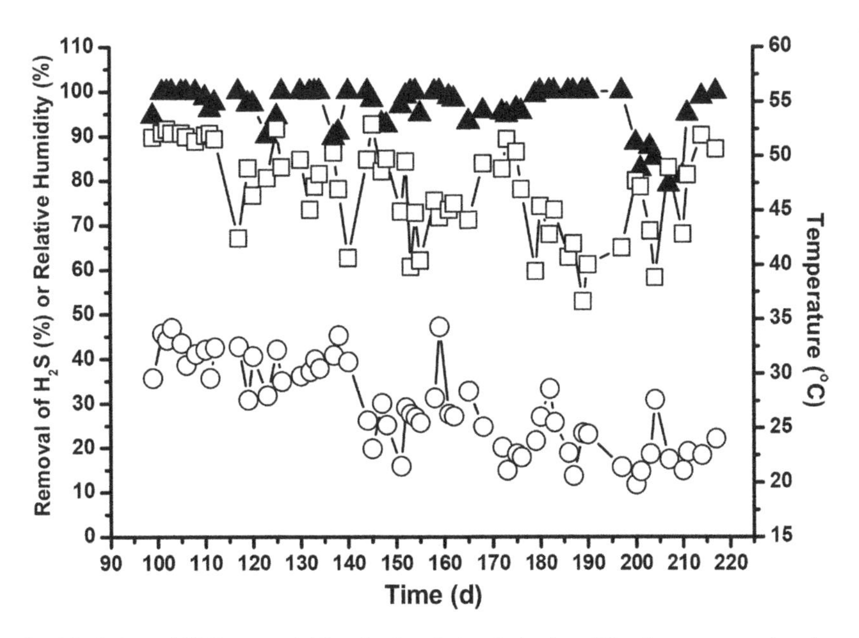

Figure 2. Variation of H_2S removal (closed triangles), relative humidity (open squares) and temperature (open circles) for flow rate of 4 litre/min (Su et al. 2014).

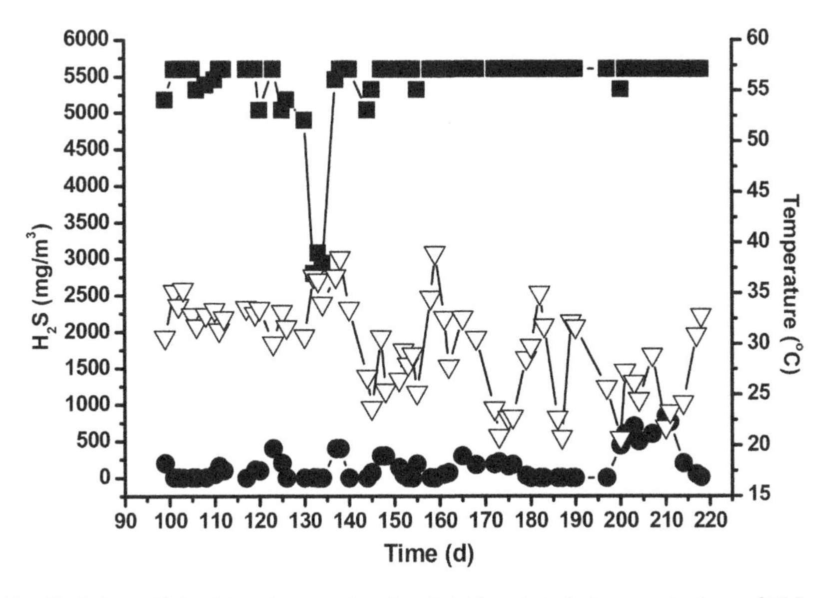

Figure 3. Variation of inlet (closed squares) and outlet (closed circles) concentrations of H_2S and inlet temperature (open inverted triangles) over time (Su et al. 2014).

or higher and the H_2S concentrations in the outlet biogas ranged within 0–560 mg/m^3 (Figure 3). Thus, 26.9 g $H_2S/m^3/h$ were removed under these conditions (5600 mg/m^3 $H_2S \times 0.24$ m^3/h ÷ 0.05 m^3 bio-carriers = 26.9 g $H_2S/m^3/h$). Under optimal operation management conditions, the efficiency of H_2S removal from the biogas was gradually stabilized.

Table 3. Soaking liquid of the bio-film coated aerial root samples from BBS by ion chromatography (Su et al. 2014).

	NH_4^+	NO_3^-	PO_4^{3-}	SO_4^{2-}
pH	mg/l			
1.8 0.47	72 3.9	0.7 0.07	587 274.1	1449 172

Data presented as mean S.D.

Seven bio-film coated aerial root samples (20 g/sample) with elemental sulfur were taken from the bio-filter and soaked in 200 ml of deionized water for 20 min. The supernatant was filtered with 0.22 μm filter discs and then analyzed for anions and cations by ion chromatography. Analytical results showed the filtered supernatant (pH = 1.8 ± 0.5) contained 1449 ± 172 mg SO_4^{2-}/L (Table 3). The low pH of the sample resulted from a high concentration of SO_4^{2-}. This implied that sulfide oxidizing bacteria had proliferated inside the bio-filter. In a recent study, sulfur oxidizing bacteria were isolated and identified from the activated sludge and bio filter bio-carriers (Su et al. 2008). In addition, an aerial root sample was randomly taken from the biogas bio-filter and analyzed for elemental sulfur by HPLC. The preliminary analytical results showed 1 g of dried aerial roots contained 0.084 ± 0.008 g of elemental sulfur.

The 6.3-L pilot-scale BBS of Su's research team can remove 26.9 g $H_2S/m^3/h$ using bio-filter approaches that operate at 7% (v/v) O_2 level continuously for more than 200 days. The experimental results showed that the efficiency of H_2S removal by a bio-filter with a mixture of aerial roots and plastic Rasching rings (H_2S removal = 90.0–100%) is better than with aerial roots only (H_2S removal = 20.0–79.2%) (Su et al. 2014).

4 DEVELOPMENT OF A FARM-SCALE BIO-DESULFURIZATION SYSTEM

This work evaluates the H_2S removal efficiency of the BBS operated on a commercial pig farm for 350 days. The optimal operating parameters for efficient H_2S removal from biogas are identified (Su et al. 2013) as well.

4.1 *The farm-scale BBS*

4.1.1 *Structure and operation of the BBS*

The farm-scale BBS was installed and operated on a commercial pig farm in Miaoli, Taiwan. The BBS comprised: (1) a biofilter (50.8 cm i.d. × 360 cm H), (2) a humidifier (30.48 cm i.d. × 70 cm H), (3) an absorption cylinder (40.6 cm i.d. × 100 cm H), and (4) a sulfur settlement basin (100 cm × 40 cm × 60 cm) (L × W × H) (Figure 4). The bio-filter unit was assembled using three identical PP columns (50.8 cm i.d. × 100 cm H each) (total working volume = 730 L) packed with bio-carriers (608 L); initial biogas flow rate was 50 L/min. Moreover, a 51-L humidifier containing 23 L clean tap water was installed, followed by the BBS to ensure that the humidity of biogas entering the BBS was higher than 85%.

The inoculum liquid (360 L) contained sludge (282 L), pond water (72 L), mud (100 g), and a sulfur-oxidizing bacterial suspension (6 L) as inoculum for the bio-filter. The inoculum liquid was re-circulated throughout the packed bio-filter by pumping the suspension via a 0.25 hp pump (TECO Co., Taiwan) from settlement basins for inoculation and then back to the settlement basins; this cycle was repeated for 24 h. The average temperature inside the bio-filter was maintained at 25–30°C. The biogas, after bio-desulfurization, was stored in bags for subsequent use. Activated sludge was added monthly (480 L/month) to the bio-filter. Piggery effluent from a piggery wastewater treatment facility was utilized to flush the

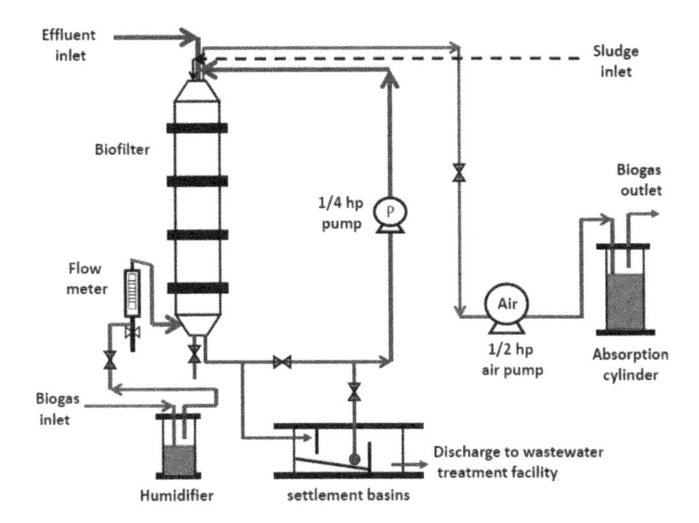

Figure 4. Sketch of the proposed farm-scale biogas bio-filter system setup (Su et al. 2013).

bio-carriers three times a week (480 L/time) in order to provide nutrients to microbes on bio-carrier surfaces.

4.1.2 *Novel operation mode by introducing adequate amount of oxygen to the BBS*
A 0.5 hp air pump (TECO Co., Taiwan) was installed at the bio-filter outlet to withdraw biogas from biogas storage bags (83.4 m^3), and introduce it into the bio-filter through a flow meter (Tohama 10B; Yeong Shin Co. Ltd., Taiwan), humidifier, and bio-filter before withdrawing the biogas using an air pump. Oxygen was essential for growing sulfur-oxidizing bacteria in the BBS and was introduced through an annulus gap in the connector of biogas inlet pipeline simply by removing an O-ring from the connector. The oxygen concentration was measured from the outlet biogas flow by a portable multi-gas leak detector (Resolution = 0.1%) (MX6 iBRID, Industrial Scientific Co., USA). The BBS operation was automatically controlled by a programmable logic controller (PLC) module.

4.2 *Novel bio-carriers for the proposed BBS*

All microorganisms for this work were isolated from a pre-pilot-scale bio-filter reactor (Su et al. 2008). The main bio-carriers were light-expanded clay aggregates (LECAs) and small Rasching rings (i.e. hollow spherical polypropylene balls) (Sheng-Fa Plastics, Inc., Taiwan). The LECA has porous matrix that can immobilize bacteria on the surface and inside the matrix. Notably, the LECA is resistant to low pH.

4.3 *Hydrogen sulfide removal by the BBS*

Overall (0–350 days) average H_2S, in the inlet and outlet biogas streams amounted to 4691 ± 1532 mg/m^3 (1231–8465 mg/m^3) and 320 ± 750 mg/m^3 (0–4001 mg/m^3), respectively (Figure 5). The start period of the BBS lasted for at least 30 days under humidity higher than 90% and mesophilic conditions for enrichment of bio-film. Operating results indicate that the average H_2S in inlet and outlet biogas were 4312.00 ± 725.37 mg/m^3 (3360–5600 mg/m^3) and 758.80 ± 870.90 mg/m^3 (0–2520 mg/m^3), respectively; the H_2S removal efficiency reached 83.33 ± 18.61% (50–100%).

During the pre-mature period (days 35–46), the average H_2S in inlet and outlet biogas were 3795.55 ± 1015.76 mg/m^3 (1680–5600 mg/m^3) and 134.30 ± 220.35 mg/m^3 (0–700 mg/m^3),

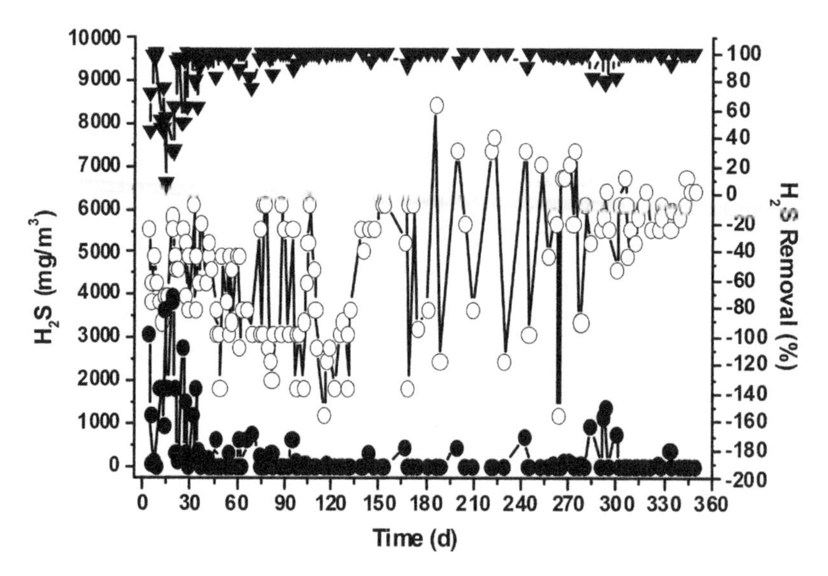

Figure 5. Variation in biogas inlet (open circles) and outlet (closed circles) H2S concentrations and H2S removal (closed inverted triangles) over time (Su et al. 2013).

respectively; the H_2S removal efficiency was 96.30 ± 6.64% (75–100%). During the stable operation period (days 47–350), the H_2S removal ranged from 96.71 ± 18.61% to 98.48 ± 3.32, under mesophilic conditions (20–37°C). Daily maximum biogas treatment volume increased to 187.2 from 72 m^3 when the biogas flow rate was raised to 130 from 50 L/min after 280 days; the average H_2S removal efficiency still exceeded 95%. The overall average H_2S removal from biogas was 93.16 ± 15.61% (7.7–100%).

4.4 *Bio-film formation and humidity build-up*

The bio-film formation period lasted for at least 35 days under adequate humidity and mesophilic conditions. Mature bio-film formation was achieved when low pH of dropping water samples was detected. Overall (0–350 days), the average humidity of inlet and outlet biogas and that of the ambient atmosphere was 92.89 ± 8.85% (66.5–99.9%), 96.27 ± 6.55% (61.6–99.9%), and 76.05 ± 14.43% (41.4–99.9%), respectively. Average humidity of inlet and outlet biogas and that of the ambient atmosphere was 84.74 ± 6.27% (74.4–94.5%), 89.64 ± 10.87% (69.5–99.9%), and 70.71 ± 15.12% (44.5–96.9%) during days 21–34, respectively (Table 4). When the biogas humidity exceeded 85%, bio-films were easily formed on the bio-carrier with higher bio-film activity.

4.5 *Sulfur oxidation by the BBS*

Analytical data showed that the average amounts of CH_4, CO_2, and N_2 in the inlet biogas equaled 60.6 ± 1.6%, 35.0 ± 1.4%, and 4.4 ± 1.5%, respectively; however, the average amounts of CH_4, CO_2, and N_2 in the outlet biogas were lower, reaching 44.7 ± 4.8%, 27.0 ± 0.7%, and 28.3 ± 5.3%, respectively (Table 5).

Roughly 5–10% air must be introduced into the bio-desulfurization unit for operation (Özmen & Aslanzadeh 2009). The oxygen (4–10%) was adequate for efficient H_2S removal in the stable operation period with biogas flow rates of 50–130 L/min (winter through summer). Notably, all CH_4 in the inlet or outlet biogas exceeded 60% (Table 5).

Table 4. Comparison of hydrogen sulfide removal efficiency at different outlet biogas temperatures and humidity percentages (Su et al. 2013).

Operation (Days)	H_2S Removal[a] (%)	Outlet biogas humidity[a] (%)	Outlet biogas temperature[a] (°C)
21–34	83.33 ± 18.61 (50–100)	89.64 ± 10.87 (69.5–99.9)	39.91 ± 4.30 (32.8–45.5)
35–46	96.30 ± 6.64 (75–100)	91.17 ± 9.31 (61.6–99.9)	34.31 ± 5.35 (32.8–47.8)
47–78	96.71 ± 6.89 (75–100)	93.24 ± 7.18 (81.2–99.9)	33.56 ± 4.97 (23.8–40.6)
78–204	98.48 ± 3.32 (84.6–100)	97.29 ± 4.78 (80.3–99.9)	25.78 ± 5.95 (14.6–36.9)
205–350	97.77 ± 5.53 (78.6–100)	98.31 ± 3.19 (87.9–99.9)	37.56 ± 5.94 (23.9–45.4)

[a]Data are mean ± SD.

Table 5. Analytical data for biogas samples by gas chromatography (Su et al. 2013).

		Components of Biogas[a] (%)		
Samples	Biogas samples	N_2	CH_4	CO_2
1	Inlet	6.1	60.1	33.7
	Outlet	24.6	48.3	27.1
2	Inlet	2.8	63.1	34.1
	Outlet	23.2	49.3	27.5
3	Inlet	4.9	59.7	35.4
	Outlet	34.3	39.8	26.0
4	Inlet	3.6	59.6	36.8
	Outlet	31.1	41.4	27.5
Mean ± SD	Inlet	4.4 ± 1.5	60.6 ± 1.6	35.0 ± 1.4
	Outlet	28.3 ± 5.3	44.7 ± 4.8	27.0 ± 0.7

[a]Mean of data in triplicates.

To maintain stable operation activity of the BBS, piggery effluent was used to flush the bio-carriers inside the bio-filter once per week. All dropping water (0.5–2.0 L/wk), flushing water, and effluent samples were periodically collected from the BBS and analyzed by means of ion chromatography and pH meters. Dropping water samples were obtained from the bottom of the bio-filter; dropping water formed by water vapor condensation inside the bio-filter. The pH value of dropping water, flushing water, and effluent was 1.54 ± 0.17, 2.25 ± 0.84, and 8.09 ± 0.18, respectively ($P < 0.01$). The average SO_4^{2-} in dropping water, flushing water, and effluent reached 27086 ± 3956, 5704 ± 2701, and 67.03 ± 16.68 mg/L, respectively ($P < 0.01$). Furthermore, the average NH_4^+ in dropping water, flushing water, and effluent was 1412 ± 488.4, 693.4 ± 359.8, and 1096.8 ± 86.3 mg/L, respectively ($P < 0.05$) by IC analysis. The average NO_3^- in dropping water, flushing water, and effluent differed significantly ($P < 0.05$).

The pH of dropping water, flushing water, and effluent was inversely proportional to the average sulfate concentration. The high COD in dropping water samples resulted from high NO^{3-} and SO_4^{2-} concentrations. The average SO_4^{2-} in dropping water samples (27086 ± 3956 mg/L) was roughly 4.7 times of that in flushing water samples (5704 ± 2701 mg/L), likely resulting in COD in dropping water samples (1456.00 ± 68.43 mg/L) roughly 4.5 times greater than in flushing water samples (322.01 ± 39.51 mg/L). Although large amounts of SS in flushing water samples originated mostly from sulfur, the amount of sulfur did not increase the BOD in flushing water samples.

A bio-scrubber composed of a gas/liquid contact tower and an aeration tank was used to remove 2800 mg/m^3 H_2S in the biogas from anaerobic digestion of potato processing wastewater (Nishimura & Yoda 1997). For full-scale operation, the H_2S removal efficiency in biogas (inlet H_2S = 7000 mg/m^3) exceeded 90% in the case of the "BIO-Sulfex" biofilter, but it is characterized by a greater power consumption for recirculating and aerating the flushing liquid (Schieder et al. 2003).

Notably, in this work, the average H_2S in the inlet biogas streams amounted to 4691 ± 1532 mg/m^3 (1231–8465 mg/m^3) and the H_2S removal efficiency reached 93–100% by only one BBS module without recirculating and aerating the flushing liquid (Su et al. 2013). In sum, the proposed BBS requires only a set of humidifier and bio-filter to remove H_2S from biogas, enabling sulfur to be recycled. Sulfur-oxidizing bacteria formed stable bio-films on bio-carrier surfaces when the dropping water samples had a pH < 2 (i.e. H_2S SO_4^{2-}). The elemental sulfur which formed in the BBS concerned in this work was automatically and periodically flushed with piggery effluent.

5 MONITORING OF SULFUR DIOXIDE EMISSION RESULTING FROM BIOGAS UTILIZIATION ON COMMERCIAL FARMS

Biogas is an aggressive gas in terms of corrosion, so the used equipment demands special care. This characteristic is a consequence of the presence of 0.1–0.5% (v/v) of H_2S (Salomon & Silva Lora 2009). The content of H_2S can be higher than 0.5% (v/v), depending on its biogas source (0.1–0.8%, v/v) (Huertas et al. 2011). Some studies imposed restrictions to allowable H_2S levels in the biogas for internal combustion engines from 15 to 150 mg/m^3 (Salomon & Silva Lora 2009; Wellinger & Lindberg 2000). Hydrogen sulfide can be removed from biogas either by non-microbial or microbial processes.

Hydrogen sulfide can be oxidized to form SO_2 ($H_2S + 1.5\ O_2 \rightarrow H_2O + SO_2$ + 518 kJ/mole) through the combustion of un-desulfurized biogas using power generators, boilers, or steam cookers (Laursen 2007). Sulfur dioxide in the atmosphere can cause acid rain through oxidation ($SO_2 + 0.5O_2 \rightarrow SO_3$ + 99 kJ/mole) and hydration ($SO_3 + H_2O \rightarrow H_2SO_4$ (gas) + 101 kJ/mole), which can pollute our living environment (Ahammad et al. 2008; Laursen 2007). Thus, both H_2S and SO_2 are toxic gases and harmful to human health.

The latest National Ambient Air Quality Standards for sulfur dioxide was set at 212 µg/m^3 in 1-h (primary standard) on June 22, 2010 (USEPA 2013). The H_2S in the biogas will depend on the feedstock composition and the digester's pH (e.g. for a manure sulfur content of 0.2% and the digester pH of 7.2, the raw biogas can contain H_2S in concentrations of nearly 3001 mg/m^3. The standard of US Occupational Safety & Health Administration (OSHA) for maximum H_2S permissible exposure level is 30 mg/m^3 in 10-min maximum duration) (UNEP 2002). The proposed and final rules for H_2S are 15 mg/m^3 as an 8-hour TWA (Time Weighted Average) and 22.5 mg/m^3 as a STEL (short-term exposure limit) (URL: http://www.cdc.gov/niosh/pel88/7783-06.html). Scrubbing the raw biogas to eliminate its H_2S and NH_3 content will prevent the formation of corrosive sulfur and nitrogen oxides, thus increasing the potential uses of the biogas.

In Taiwan, more than 95% of commercial pig farms (total 5.8 million pigs on farms in 2014) are equipped with underground, horizontal anaerobic digesters and wastewater treatment facilities. However, un-desulfurized biogas is applied to heating lamps of piglets in winter and is emitted into the atmosphere in summer. Biogas utilization was idled for 20 years in Taiwan until successful development of biogas bio-filter facility to remove H2S in livestock biogas (Su et al. 2008; 2013; 2014). However, most pig farmers still do not understand the essence of biogas desulfurization for human health and environmental concerns. The main goal of this study was to evaluate the H_2S and SO_2 emission in exhaust gases and the human health issue after un-desulfurized biogas combustion.

5.1 *Commercial pig farms*

Three commercial pig farms equipped with power generators, boilers or hot water stoves were selected for this study (Table 6). Un-desulfurized biogas was used as the sole fuel for power generators, boilers and hot water stoves. Pig farm (I) (1000 pigs on farm) has been equipped with a set of biogas bio-filter facilities for desulfurizing biogas (Su et al. 2013; 2014). Pig farm (II) (200 pigs on farm) has been equipped with a boiler. Pig farm (III) (300 sows on farm) has been equipped with a hot water stove.

For Pig farm (I), the biogas mixture consisted of desulfurized and un-desulfurized biogas by the ratio of 4:1, resulting from some portion of the un-desulfurized biogas that had entered the desulfurized gas pipelines. All exhaust gas samples were detected for H_2S and SO_2 concentrations on site.

5.2 *Monitoring the exhaust gas from biogas combustion*

In addition, other commercial pig farms equipped with power generators were also selected for monitoring H_2S and SO_2 concentrations in the exhaust gas samples by using either desulfurized or un-desulfurized biogas as the sole fuel. A stainless steel sampling tubing was connected to all gas detectors (PortaSens II detector for H_2S; MX6 iBRID detector for SO_2) for reducing the exhaust gas temperature through Teflon tubing. The sketch of experiment design was shown in Figure 6. Biogas produced from anaerobic digesters of pig farm was introduced to either a bio-filter facility by a biogas vacuum pump or the power generator. The parameters for biogas production and application forms of three pig farms are shown in Table 6.

5.3 *Evaluation of hydrogen sulfide and sulfur dioxide emission from biogas combustion*

For Pig farm (I), the concentrations of H_2S and SO_2 were 5575 ± 862 mg m³ and 19.2 ± 5.6 mg/m³ in un-desulfurized biogas, respectively (Table 7). Combustion of un-desulfurized biogas can produce sulfur dioxide ($2H_2S + 3O_2 \rightarrow 2H_2O + 2SO_2$). When the un-desulfurized biogas was used for the power generator, the concentrations of H_2S and SO_2 reached 113 ± 43 and 256 ± 113 mg/m³ in exhaust gas of power generator, respectively. Analytical results of biogas combustion showed that the combustion of un-desulfurized biogas exhausted about 92.5%

Table 6. Parameters for biogas production of three commercial pig farms (Su and Chen 2015).

Pig farm	Biogas source	Daily wastewater volume (m³/d)	Biogas yield* (m³/d)	Biogas flow rate (L/min)	H_2S in biogas (mg/m³)	Biogas utilization (specifications)
I	Anaerobic digestion of piggery wastewater	30	100	50 for biofilter	5.575 ± 862	Power generator (30 kW)
II		6	20	NA	2.510 ± 1061	Boiler (5.7 kg-biogas/h)
III		10	30	NA	4.562 ± 1298	Hot water stove (1.7 kg-biogas/h)

**Pig number multiples 0.1 m³ per pig per day measured parameter on site.*
NA: Not available.

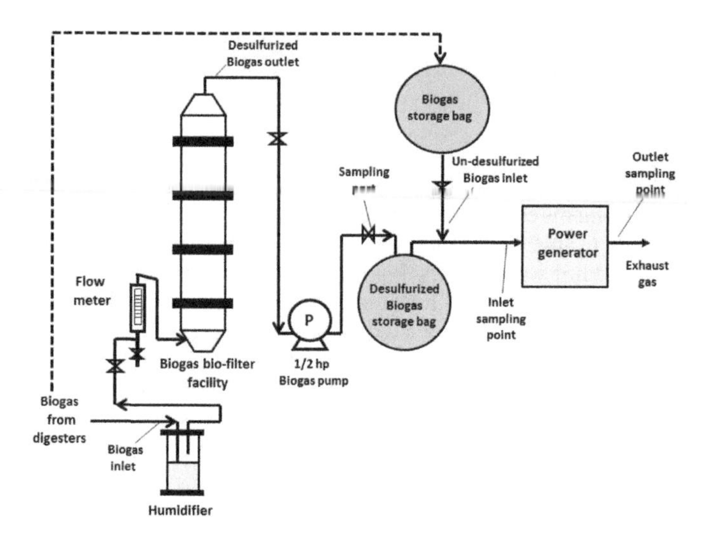

Figure 6. The sketch of the experimental design for biogas bio-filter facility and power generator (Su and Chen 2015).

more SO_2 ($P < 0.01$). In the meantime, about 98% of hydrogen sulfide was removed during the combustion process using un-desulfurized biogas ($P < 0.01$). Hydrogen sulfide can easily react with metal parts of the power generator and reduce the H_2S concentrations in the exhaust gas. Experimental results showed that 95.4% of H_2S were converted to SO_2 and most SO_2 seemed to deposit on the surface of the internal combustion engine (Table 7). This study was only focused on monitoring the H_2S and SO_2 emissions from biogas combustion not including H_2SO_4 in liquid phase (i.e. $SO_2 + 1/2O_2 \rightarrow SO_3$; $SO_3 + H_2O \rightarrow H_2SO_4$), thus, SO_3 was not considered.

The concentrations of H_2S and SO_2 in the desulfurized biogas were 3.3 ± 1.8 and 0.3 ± 0.6 mg/m^3, respectively (Table 7). The mixed biogas resulted from the biogas recirculation design to avoid over suction pressure of biogas vacuum pump and prevent vacuum phenomenon inside the biogas bio-filter facility. When the biogas mixture (desulfurized: un-desulfurized = 4:1; v:v) was used for the power generator, the concentrations of H_2S and SO_2 in exhaust gas of the power generator were 19.8 ± 19.1 and 52.7 ± 53.6 mg/m^3, respectively. The analytical results of biogas combustion showed that the combustion of desulfurized biogas mixture exhausted about 99.4% more SO_2 ($P < 0.01$). In the meantime, about 83.3% of hydrogen sulfide was removed during the combustion process using desulfurized biogas mixture. Because exhaust gas from the power generator was directly detected on farm in an open environment, data variation detected from the exhaust gas was relatively high.

For Pig farm (II), the concentrations of H_2S and SO_2 in the un-desulfurized biogas were $2{,}510 \pm 1061$ and 13.3 ± 7.3 mg/m^3, respectively. When the un-desulfurized biogas was used for a boiler, the concentrations of H_2S and SO_2 were in the exhaust gas of the boiler 93 ± 73.4 and 208 ± 82.4 mg/m^3 respectively (Table 8). The analytical results of biogas combustion showed that the combustion of un-desulfurized biogas exhausted about 93.6% more SO_2 ($P < 0.01$). In the meantime, about 96.6% of hydrogen sulfide was removed during the combustion process using un-desulfurized biogas ($P < 0.01$). The experimental results showed that 91.7% of H_2S were converted to SO_2 and most SO_2 seemed to deposit inside the boiler (Table 8).

For Pig farm (III), the concentrations of H_2S and SO_2 in the un-desulfurized biogas were $4{,}562 \pm 1298$ and 15.8 ± 9.87 mg/m^3, respectively. When the un-desulfurised biogas was used for a hot water stove, the concentrations of H_2S and SO_2 in the exhaust gas of the hot water stove were 34.8 ± 19.2 and 51.1 ± 40.3 mg/m^3, respectively (Table 9). Analytical results of biogas combustion showed that the combustion of un-desulfurized biogas exhausted about 69.1% more SO_2 ($P < 0.05$). In the meantime, about 99.2% of hydrogen sulfide was removed during the combustion process using un-desulfurized biogas ($P < 0.01$). The experimental results showed that 90.9% of H_2S was converted to SO_2 and most SO_2 seemed to deposit on the hot water stove (Table 9).

To date, no hydrogen sulfide or sulfur dioxide regulatory emission limits have been set for livestock farming in Taiwan. Biogas produced from pig farms is released into the atmosphere in the summer. Some un-desulfurized biogas is used for heating lamps for piglets in the winter. The combustion of un-desulfurized biogas can produce sulfur dioxide and can cause air pollution problems. More than 96% of hydrogen sulfide in the biogas was reduced in the case of three commercial pig farms which used it for power generation or direct combustion using boilers and hot water stoves (Tables 7 to 9). Thus, the best way to control the hydrogen sulfide and sulfur dioxide emission is to lower the hydrogen sulfide amount in the livestock biogas before any form of biogas combustion is carried out.

Table 7. Detection of H_2S and SO_2 in gas samples using un-desulfurised biogas for power generation on Pig farm (I) (Su and Chen 2015).

Un-desulfurised biogas data (n = 18)					
Concentration (mg/m^3)	Biogas	Exhaust gas	Difference (%)	*P*	H_2S conversion degree (%)
H_2S	5575 ± 862	113 ± 43	−98.0	< 0.01	95.4
SO_2	19.2 ± 5.6	256 ± 113	+92.5	< 0.01	
Desulfurised biogas mixture data (n = 18)					
Concentration (mg/m^3)	Biogas	Exhaust gas	Difference (%)	*P*	H_2S conversion degree (%)
H_2S	3.3 ± 1.8	19.8 ± 19.1	+83.3	NS	NA
SO_2	0.3 ± 0.6	52.7 ± 53.6	+99.4	< 0.01	

*Using desulfurised biogas mixture (desulfurised: un-desulfurised = 4:1; v:v) for power generation
n, sample size, Data presented as means ± S.D. NS, not significant NA, not applicable.

Table 8. Detection of H_2S and SO_2 in gas samples using un-desulfurised biogas for combustion on Pig farm (II) (Su and Chen 2015).

Concentration (mg/m^3)	Biogas (n = 11)	Exhaust gas (n = 11)	Difference (%)	*P*	H_2S conversion degree (%)
H_2S	2,510 ± 1061	93 ± 73.4	−96.6	< 0.01	91.7
SO_2	13.3 ± 7.3	208 ± 82.4	+93.6	< 0.01	

n, sample size.
Data presented as means ± S.D.

Table 9. Detection of H_2S and SO_2 in gas samples using un-desulfurised biogas for combustion on Pig farm (III) (Su and Chen 2015).

Concentration (mg/m^3)	Biogas (n = 10)	Exhaust gas (n = 10)	Difference (%)	*P*	H_2S conversion degree (%)
H_2S	4562 ± 1298	34.8 ± 19.2	−99.2	< 0.01	90.9
SO_2	15.8 ± 9.87	51.1 ± 40.3	+69.1	< 0.05	

n, sample size.
Data presented as means ± S.D.

6 DEMONSTRATION OF FARM SCALE BIO-DESULFURIZATION SYSTEM FOR LIVESTOCK BIOGAS IN TAIWAN

Biogas contains 2800 + 8400 mg/m^3 of hydrogen sulfide, produced as a by-product of anaerobic digestion of organic matter (e.g. animal manure, sewage sludge, and food-processing waste), fermentation of animal manure and other wastes (Syed et al. 2006; Su et al. 2013). Combustion of biogas containing hydrogen sulfide yields sulfur oxides, thus limiting the use of biogas for heat and power generation. Importantly, sulfide and sulfur dioxide corrode metal generator parts and contaminate the engine oil in biogas motors. Additionally, most industrial factories operate a conventional water scrubber system to remove hydrogen sulfide from biogas. This system requires considerable amounts of water and electricity to dissolve hydrogen sulfide in water, thus increasing the operating costs but making the reduction of hydrogen sulfide concentrations in biogas feasible.

However, a novel and cost-effective biogas biofilter system (BBS) for bio-desulfurization (H_2S S^0 SO_4^{2-}) was successfully developed by our research team and is being used in more than four pig farms (Su et al. 2008; 2013; 2014; 2015) (Figure 7). Both pilot and full-scale BBS were tested and evaluated on selected pig farms from 2006 to 2010, during which time the average hydrogen sulfide removal efficiency exceeded 95% (Su et al. 2013; 2014) (Figure 8).

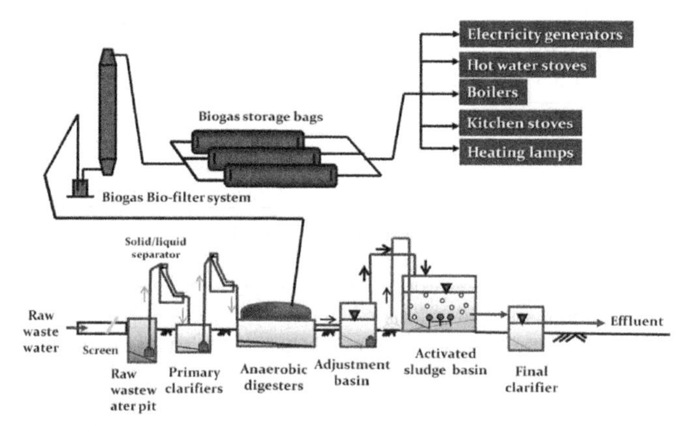

Figure 7. Flowchart of the piggery wastewater treatment system with a biogas desulfurization biofilter (Su, 2015).

Figure 8. Farm-scale biogas biofilter system established in 2009 (for 9,000-head pig farm).

This farm-scale BBS with three patents was conferred an award at the 7th National Innovation Award in Taiwan in 2010.

6.1 *Demonstration of biogas power generation and clean development mechanism (CDM) practice on a 9000-head pig farm*

The farm-scale BBS was installed and operated on a 9000-head pig farm, the Ping-Shun Pig Farm, Miaoli, Taiwan. Approximately 900 m^3 biogas, containing 60% methane is produced daily. One set of 30 kW power generators is installed and operated daily. Thus, 240 kWh of electricity is produced on a daily basis. The BBS comprised: (1) a biofilter (71 cm i.d. × 360 cm H × 3 sets), (2) humidifier (30.48 cm i.d. × 70 cm H), (3) absorption cylinder (40.6 cm i.d. × 100 cm H), and (4) sulfur settlement basin (100 cm × 40 cm × 60 cm) (L × W × H).

The livestock industry can obtain voluntary carbon units (VCUs) from biogas utilization after complying with voluntary carbon standard (VCS) criteria. The manufacturing industry can create joint ventures by combining carbon-offset programs with the livestock industry for biogas utilizations on farms. The first document of VCS's project design document (PDD) was completed mainly for a 9000-pig farm (Ping-Shun pig farm) in 2009. Two methodologies (AMS-III.H. and AMS-I.D.) of the Clean Development Mechanism (CDM) were used for PDD preparation. The validated PDD indicates that approximately 3163 tCO_2e per year can be reduced by biogas electricity production for a 9,000-pig farm with only one set of 30 kW power generators. However, the biogas volume for a 9000-pig farm can provide three sets of 30 kW power generators. About 9489 tCO_2e per year can be reduced by biogas electricity production for a 9000-pig farm with three 30 kW power generators. This finding implies that the combustion of biogas produced from one pig's manure can be counted as 1 tCO_2 equivalent per year (i.e. 1 tCO_2e/pig/yr). Moreover, approximately 6195 $ktCO_2e$ per year can be reduced when biogas is used for power generation or other direct combustion. The first copy of the validation report (ISO 14064-2) and statement of carbon reduction for a pig farm was validated by SGS Taiwan, Ltd. with the BBS system (Figure 9).

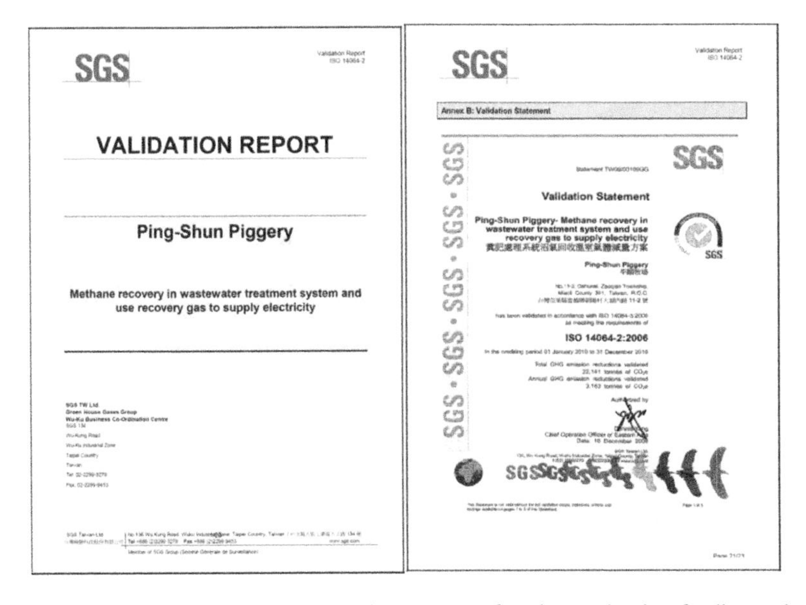
SGS

Validation Report
ISO 14064-2

VALIDATION REPORT

Ping-Shun Piggery

Methane recovery in wastewater treatment system and use recovery gas to supply electricity

SGS TW Ltd
Green House Gases Group
Wu-Ku Business Co-Ordination Centre

SGS

Validation Report
ISO 14064-2

Annex B: Validation Statement

Validation Statement

Ping-Shun Piggery- Methane recovery in wastewater treatment system and use recovery gas to supply electricity

Ping-Shun Piggery

ISO 14064-2:2006

Figure 9. First copy of validation report and statement of carbon reduction for livestock farms in Taiwan (Su, 2015).

6.2 *Demonstration of biogas power generation on a 25,000-head pig farm*

Another farm-scale BBS was established and put in operation in 2013 at a 25000-head pig farm, the Central Pig Farm, Pingtung County, Taiwan (Figure 10). The 25000-head pig farm has been selected as a demonstration site for biogas power generation in 2015 by Bureau of Energy, Ministry of Economic Affairs, Taiwan (Figure 11). About 2500 m^3 of biogas containing 60% methane is produced daily. Three sets of 65 kW microturbine power generators are installed and operated sequentially. Thus, an average of 3700 kWh electricity is produced daily.

Figure 10. Contract signing press conference for biogas power generation with a farm-scale biogas biofilter in June 20, 2013 (for 25,000-head pig farm) (Su, 2015).

Figure 11. Contract signing press conference for biogas power generation with a farm-scale biogas biofilter in December 30, 2013 (for 25,000-head pig farm) (Su, 2015).

7 CONCLUSIONS

Livestock biogas collection and utilization is gaining increasing attention in Taiwan, along with other renewable energy such as solar and wind power. Taiwan Council of Agriculture (TCOA) and Environmental Protection Agency (TEPA) are encouraging livestock farmers to promote the efficiency of their anaerobic digesters and perform biogas power generation on farm. Thus, the digestate from anaerobic digestion of wastewaters from dairy and pig farms can be applied to agricultural lands as organic fertilizers through legal application procedure and authorized by TEPA since September 22, 2015 (http://enews.epa.gov.tw/enews/fact_Newsdetail.asp?InputTime=1040922160957).

In addition, the livestock farms performing biogas collection and power generation do not need to pay for the water pollution control fee after January 1st, 2017. The patent of BBS technology had been non-exclusive technology licensed to Shinesun Bio-energy Inc., Ltd. and Sharp-Ray Energy Inc., Ltd., in 2009 and 2011, respectively. The BBS technology can be employed in coordination with biogas upgrading system for the production of bio-natural gas and applied to fuel cell, vehicle fuel, and natural gas grid. Biogas absorption chiller will be developed and demonstrated on livestock farms with BBS facilities in the near future because of humid and warm weather in Taiwan.

ACKNOWLEDGEMENTS

The study was made possible by grants awarded from the National Science Technology Program for Agriculture Biotechnology (NSTPAB), the Council of Agriculture (COA), (Project No. 98 AS-1.2.1-AD-U1; No. 99 AS-7.2.1-AD-U1), and the National Science Council (NSC) (Project No. 95C-0803; No. NSC 97-2317-B-059-005), Executive Yuan, Taiwan.

REFERENCES

Ahammad, S.Z., Gomes, J., Sreekrishnan, T.R. 2008. Wastewater treatment for production of H2S-free biogas. *Journal of Chemical Technology & Biotechnology*, 83: 1163–1169.

Basu, R., Clausen, E.C., Gaddy, J.L. 1996. Biological conversion of hydrogen sulfide into elemental sulfur. *Environ. Progress* 15(4): 234–238.

Cao, Y., You, J., Wang, R., Shi, Y. 2016. Designing a Mixed Evaluating System for Green Manufacturing of Automotive Industry. *Problemy Ekorozwoju/Problems of Sustainable Development*, 11(1): 73–86.

Cel, W., Czechowska-Kosacka, A., Zhang, T. 2016. Sustainable Mitigation of Greenhouse Gases Emissions. *Problemy Ekorozwoju/Problems of Sustainable Development*, 11(1): 173–176.

Craggs, R., Park, J., Heubeck, S. 2008. Methane emission from anaerobic ponds on a piggery and a dairy farm in New Zealand, *Australian Journal of Experimental Agriculture*, 48: 142–146.

Huertas, J.I., Giraldo, N., Izquierdo, S. 2011. *Removal of H2S and CO2 from Biogas by Amine Absorption. Mass Transfer in Chemical Engineering Processes*, edited by Jozef Markoš (ISBN 978-953-307-619-5), InTech, pp. 133–151. (URL: http://cdn.intechopen.com/pdfs/22869/InTech-Removal_of_h2s_and_co2_from_biogas_by_amine_absorption.pdf).

IPCC 2007. Climate Change 2007: The Physical Science Basis. Contribution of Working Group I to the Fourth Assessment Report of the Intergovernmental Panel on Climate Change (Eds S. Solomon, D. Qin, M. Manning, Z. Chen, M. Marquis, K.B. Averyt, M. Tignor & H.L. Miller). Cambridge, UK: Cambridge University Press.

Janssen, A.J.H., Sleyster, R., Kaa, C.V.D., Jochemsen, A., Bontsema, J., Lettinga, G. 1995. Biological sulphide oxidation in a fed-batch reactor. *Biotech. Bioengineer*, 47: 327–333.

Kantachote, D., Charernjiratrakul, W., Noparatnaraporn, N., Oda, K. 2008. Selection of sulfur oxidizing bacterium for sulfide removal in sulfate rich wastewater to enhance biogas production, *Electronic Journal of Biotechnology*, 11 (2). Available from: http://www.ejbiotechnology.info/content/vol11/issue2/full/13/13.pdf (accessed 12 March, 2012).

Kapdi, S.S., Vijay, V.K., Rajesh, S.K., Prasad, R. 2005. Biogas scrubbing, compression and storage: perspective and prospectus in Indian context. *Renewable Energy* 30: 1195–1202.

Kim, Y.J., Kim, B.W., Chang, H.N. 1996. Desulfurization in a plate-type gas-lift photobioreactor using light emitting diodes. *Korean J. Chem. Engineer*, 13(6): 606–611.

Laursen, J.K., 2007. The process principles of sulfur recovery by the WSA process. *Hydrocarbon Engineering*. (http://www.topsoe.com/business_areas/gasification_based/~/media/PDF%20files/WSA/Topsoe_%20WSA_process_principles.ashx).

Lieffering, M., Newton, P., Thiele, J.H. 2008. Greenhouse gas and energy balance of dairy farms using unutilised pasture co-digested with effluent for biogas production. *Australian Journal of Experimental Agriculture* 48:104–108.

Moosavi, G.R., Mesdaghinia, A.R., Naddafi, K., Vaezi, F., Nabizadeh, R. 2005. Biotechnology advances in treatment of air streams containing H2S, *J. Biol. Sci.* 5(2): 170–175.

Murphy, J.D., McCarthy, K. 2005. The optimal production of biogas for use as a transport fuel in Ireland. *Renewable Energy 30*: 2111–2127.

Nishimura, S., Yoda, M., 1997. Removal of hydrogen sulfide from an anaerobic biogas using a bioscrubber. *Water Sci. Technol.* 36(6–7): 349–356.

Omer, A.M. & Fadalla, Y. 2003. Biogas energy technology in Sudan. *Renewable Energy* 28: 499–507.

Özmen, P. & Aslanzadeh, S. 2009. Biogas production from municipal waste mixed with different portions of orange peel, Master's thesis of School of Engineering, University of Borås. (http://www.susane.info/en/ref/Aslanzadeh,%20%C3%96zmen.pdf).

Potivichayanon, S., Pokethitiyook, P., Kruatrachue, M. 2005. Hydrogen sulfide removal by a novel fixed-film bioscrubber system. *Process Biochem.* 41: 708–715.

Prasertsan, S., Sajjakulnukit, B. 2006. Biomass and biogas energy in Thailand: potential, opportunity and barriers. *Renewable Energy* 31: 599–610.

Salomon, K.R. & Silva Lora, E.E. 2009. Estimate of the electric energy generating potential for different sources of biogas in Brazil. *Biomass and Bioenergy* 33: 1101–1107.

Schieder, D., Quicker, P., Schneider, R., Winter, H., Prechtl, S., Faulstich, M. 2003. Microbiological removal of hydrogen sulfide from biogas by means of a separate biofilter system: experience with technical operation. *Water Science and Technology* 48: 209–212.

Su, J.J. & Chen, Y.J. 2015. Monitoring of sulfur dioxide emission resulting from biogas utilization on commercial pig farms in Taiwan. *Environmental Monitoring and Assessment* 187(1): 4109 (total 8 pages).

Su, J.J., Chen, Y.J., Chang, Y.C. 2014. A study of a pilot-scale biogas bio-filter system for utilization on pig farms. *Journal of Agricultural Science* 152(2): 217–224.

Su, J.J., Chang, Y.C., Chen, Y.J., Chang, C.K., Lee, S.Y. 2013. Hydrogen sulfide removal from livestock biogas by a farm-scale bio-filter desulfurization system. *Water Science and Technology*, 67(6): 1288–1293.

Su, J.J., Chen, Y.J., Chang, Y.C., Tang, S.C. 2008. Isolation of sulfur oxidizers for desulfurizing biogas produced from anaerobic piggery wastewater treatment in Taiwan. *Australian Journal of Experimental Agriculture*, 48(1–2):193–197.

Su, J.J., Liu, B.Y., Chang, Y.C. 2003. Emission of greenhouse gas from livestock waste and wastewater treatment in Taiwan. Agriculture. *Ecosystems & Environment* 95: 253–263.

Syed, M., Soreanu, G., Falletta, P., Béland, M. 2006. Removal of hydrogen sulfide from gas streams using biological processes: a review. *Canadian Biosystems Engineering*, 48: 2.1–2.14.

UNEP, 2002. Guidelines for Biogas systems (Release 1.0), Environmental Due Diligence (EDD) of Renewable Energy Projects, United Nations Environment Program (UNEP) (URL: http://www.energy-base.org/uploads/media/EDD_Biogas_Systems.pdf).

USEPA, 2013. Sulfur Dioxide (SO_2) Primary Standards: Table of History of the National Ambient Air Quality Standard for Sulfur Dioxide during 1971–2010. US Envornmental Protection Agency (USEPA) (URL: http://www.epa.gov/ttn/naaqs/ standards/so2/s_so2_history.html)

Velkin, V.I. & Shcheklein, S.E. 2017. Influence of RES Integrated Systems on Energy Supply Improvement and Risks. *Problemy Ekorozwoju/Problems of Sustainable Development*, 12(1): 123–129.

Wellinger, A. & Lindberg, A. 2000. Biogas upgrading and utilization. Task 24: Energy from biological conversion of organic waste, *IEA Bioenergy*, pp.1–19.

New concept of biogas system as renewable energy and multi-generation systems for sustainable agriculture-acceleration or selection of biochemical reaction

J. Takahashi
Obihiro University of Agriculture and Veterinary Medicine, Obihiro, Hokkaido, Japan

ABSTRACT: The newly developed thermophilic fermentation system, which operates at a higher temperature than the conventional one, such as 60° or 65°, enables to reduce the size of a structure by miniaturization of digesters due to their shorter retention times, easy excess ammonia stripping from digested slurries as well as the pasteurization or inactivation of bacteria, viruses and seeds of weeds. Various agricultural wastes could be co-digested fluently by hyper-thermophilic or thermophilic fermentations after an effective degradation of biomasses as the pretreatment. It is also possible for biogas systems to reduce the generation costs by scaling up of the systems from small individual to large centralized plants, which also enables to obtain higher efficiency in power generation. Biogas plants are in an advantageous position in regard to the generation cost, compared to photovoltaic and wind turbine generations, since it is not necessary for biogas generation to provide batteries for energy storage. Additionally, ammonia stripping technology may provide a strategic procedure to eliminate compounds rich in inorganic nitrogen from the digested slurry. The quantitative reduction of nitrogen in the effluent might contribute to the mitigation of ammonia and nitrous oxide (N_2O) emission in the atmosphere, and nitrate in the hydrosphere. Recycling strategies of the stripped nitrogen from digested slurry using ammonia stripping apparatus as an annexable facility to biogas plant is proposed. Furthermore, the possible control of redox potential was applied to modulate methanogenesis and bio-desulfurization by newly developed potential controlled bio-reactor.

1 INTRODUCTION

According to an IPCC scenario (2000), biomass energies including biogas systems are characterized by "renewable, abundant, carbon neutral, storable and substitutive" as key words, and are expected to supply 1/3 of global energy consumption at the end of this century. Biogas plant has attracted much attention recently from the viewpoints of bio-waste recycling for environmental preservation and a mitigating option of greenhouse gases as one of the renewable energies (Sasse 1988, Mao 2015, Cel et al. 2016, Żukowska et al. 2016). However, biomass energies still have a major obstacle to their wide use due to relatively lower cost performance compared with large scale wind turbine systems, problems in epidemiology and nitrogen pollution in hydrosphere resulting from nitrate nitrogen contamination, as well as ammonia and nitrous oxide emission into the atmosphere from spraying the digested biomass slurries after anaerobic fermentation (Puckett et al. 1999, Dai et al. 2015). Nevertheless, biogas plants are an acceptable solution for renewable energy supply as CHP. Combined heat and power) plants, epidemiological problem and environmental preservation by mitigation of GHG (methane and nitrous oxide) due to disposal of organic wastes, sanitary reuse of the digested organic wastes should be given the first priority in consideration (Kebreab, et al., 2006). Thus, the type of fermentation should be thermophilic, 55°, or hyper-thermophilic, such as 65° (Takahashi 2010, Moset et al. 2015). The value of biogas as power generation source has been amplified by desulfurization. Bio-desulfurization has been established as low-input sustainable method.

The present paper deals with the new technology of hyper-thermophilic fermentation and possible control of redox potential to modulate methanogenesis and bio-desulfurization in a newly developed potential controlled bio-reactor.

2 THE ADVANCED BIOGAS SYSTEM PERFORMANCE

Figure 1 shows a typical advanced biogas system design, which is characterized by the following items; (1) thermophilic and hyper-thermophilic digestion makes qualitative improvements of digested slurries in sterilizing sanitation, nitrogen concentration reduction, and prevention of foul odor emission, (2) stable supply and increased calorific value of the biogas in this system enables the gas engine co-generation to be superior in regard to the total generation costs, including the energy storage compared with photovoltaic or wind turbine application, and (3) recycling of stripped nitrogen from digested slurry nitrogen using ammonia stripping apparatus as a facility annexable to a biogas plant. Figure 2 shows sterilizing conditions for animal viruses and a fecal bacterium. Most of pathogens which can be contaminated in the livestock wastes are killed off by thermophilic and hyper-thermophilic digestion. Table 1 shows ammonia and foul odor removal capacity of thermophilic and hyper-thermophilic demonstration plants. An increase in the digestion temperature can improve the ammonia stripping efficiency; accompanying odorous gases such as hydrogen sulfide (H_2S) and methyl mercaptan (MeSH) can be eliminated at over 55°C.

Digested slurries of biogas plants involve thousands of milligrams of nitrogen in the form of ammonia and organic amino compounds, which causes contamination of underground water by nitrate ions and a GHG emission of N_2O from spraying them on the fields. Table 2 shows some examples of nitrous oxide emission in the case of digested cattle manure sprayed to pastures. The emission rates greatly depend on the redox conditions and the moisture contents of the soil sprayed by digested slurries. Since it seems difficult to constantly avoid the condition of soil emitting the nitrous oxide, the best way to control the nitrous oxide emission is to reduce the nitrogen content reduction in digested slurries.

Table 3 shows the comparison of generation costs per kWh in renewable energies. For energy storage functions in biogas systems, it is possible for biogas engines to obtain the high efficiency of electric power generation in natural gas and biogas generations, compared with gas turbine generations. The efficiencies of electric power generations are almost 50% for MW class gas engines and over 30% for 20–30 kW class engines, in contrast to about 25% for

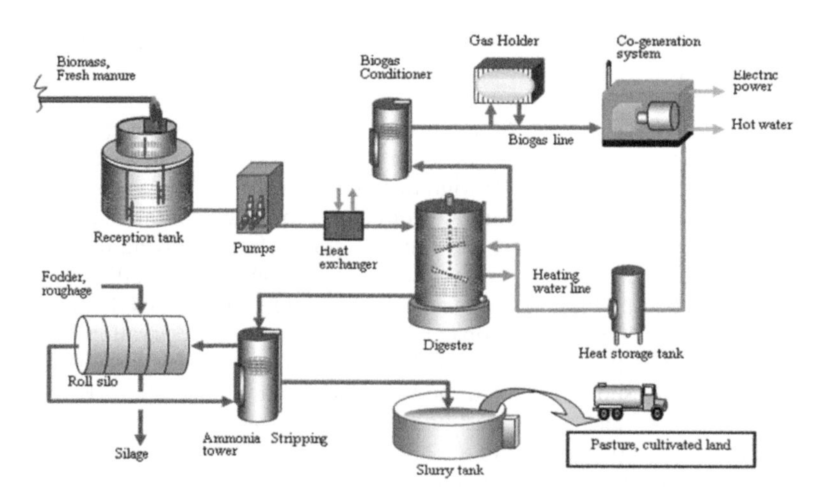

Figure 1. An advanced biogas system as a CHP plant for sustainable agriculture (Takahashi, 2011).

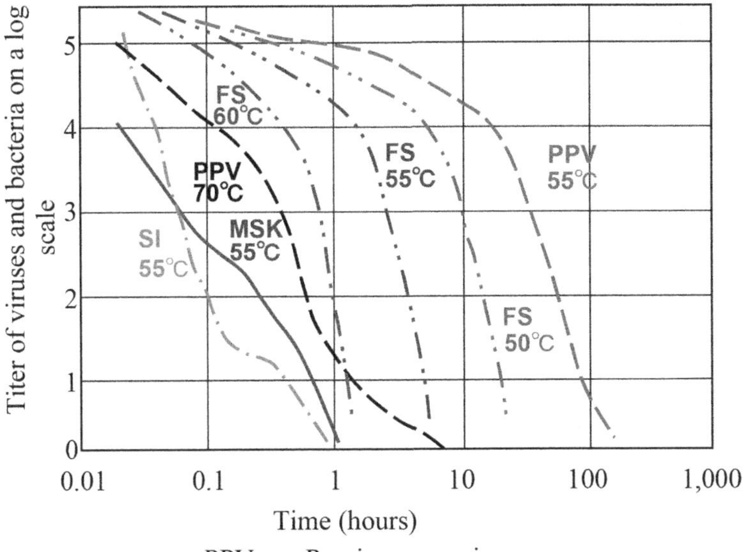

PPV : Porcine parvovirus,
MSK : Foot-and mouth disease virus,
SI : Swine influenza virus and
FS : Faeces entrococci.

Figure 2. Inactivation of animal viruses and a fecal bacterium at various temperatures (Benedixen, 1997).

Table 1. Ammonia stripping performance and odorous gas concentrations in the stripped gases.

	NH_3-N concentration in digested manure		Concentration in the stripped gases		
Digested manure	Before NH_3 stripping (mg/ℓ)	After NH_3 stripping (mg/ℓ)	NH_3 (ppm)	H_2S (ppm)	MeSH (ppm)
Mesophilic pH 7.8 (approx. 37°C)	2110	1950	155	250	26
Thermophilic pH 7.5 (approx. 53°C)	1650	1140	470	30–87	0–5
Hyper-thermophilic pH 7.5 (approx. 62°C)	1870	380	970	12	0

MW class gas turbines. From the view point of demand side conformity, however, it is possible for natural gas and biogas generation systems to store energies in the form of gaseous condition, which is the same function as the battery energy storage for load leveling, peak shaving and smart grid applications. The energy storage of biogas systems by gas holders is to superior to battery storages in terms of construction costs, whereas photovoltaic and wind turbine generations need batteries for their effective energy use. Comparing the total generation costs of biogas systems with those of other renewable energy systems, such as photovoltaic panels and wind turbines, especially in the case of standalone types of biogas storage by gas holders, gives them an advantage of inexpensive generation cost, since photovoltaic panels and wind turbines need relatively expensive batteries for the storage of the

Table 2. N_2O emissions from cattle manure treatments.

Sample no.	Sampling Spots in the treatment processes	pH	Redox potentials measured in the lab. (mV vs. Ag/AgCl)	Dry matter (wt%)	600° ignition losses (wt%)	Nitrogen contents T-N (mg/ℓ)	NH_3-N (mg/ℓ)	NO_2^--N (mg/ℓ)	NO_3^--N (mg/ℓ)
1	Covered pond	6.8	–98	6.35	2.42	471	111	0.20	30.0
2	Open pond	7.3	–71	8.93	5.43	283	21.0	0.01	17.5
3	Lagoon	7.1	–45	14.4	4.77	620	81.0	1.00	22.5
4	Cattle manure pond	6.5	–72	10.7	3.87	140	17.0	0.00	7.50
5	Obihiro cattle Digested manure	7.8	–172	3.98	3.10	1510	1480	0.70	20.0

Sample no.	Gas generated: Methane concentration ppm	Nitrous oxide concentration ppb	N_2O evolution in the case of beef cattle manure treatments kg of N_2O/(a beef cattle · Year)	Remarks (Estimated gas evolution rates: approx. 300 ml/g-VS, 20°)	
1	0	205,000	0.240	995 ppm of CH_4 and 3,600 ppb of N_2O were produced at –156 mV of redox potential	Australia
2	0	395,000	0.460	920 ppm of CH_4 and 4,100 ppb of N_2O were produced at –190 mV of redox potential	Australia
3	0	6,000	0.007	–	Australia
4	0	32,000	0.037	–	Australia
5	440	8,000	0.009	–	Japan

(Takahashi, 2008).

Table 3. Comparison of generation costs per kWh.

Biogas plants with biomass transportation, biomass pre-treatment, digested slurry treatment and co-generation facilities	US¢ 30–80
PV generation system with BESS	US¢ 60–100
WT generation system with BESS	US¢ 50–100

BESS: Battery energy storage system.
(Stationary type lead-acid battery).

generated electricity. The energy storage method by gas holder makes supply and demand adjustments easy in the case of grid connected power systems. Ammonia stripping technology may provide a strategic valid procedure to eliminate rich inorganic nitrogen content in the digested slurry.

3 INTRODUCING POTENTIAL CONTROLLED BIO-REACTORS

To further improve the biogas system performance, introducing potential controlled bio-reactor was attempted. The bio-reactor makes it possible to control oxidation-reduction (redox) potential as one of the most effective factors connected directly with free energy changes of the considered bio-reactions. Figure 3 shows the pH-potential diagram related to carbon and nitrogen. The potential controlled bioreactors can modulate microbial or enzymatic redox and hydrolysis reactions. Possible involvements to redox reactions comprise acceleration of methane fermentation, hydrogen fermentation selectivity, lactic acid fermentation selectivity, mitigation of nitrous oxide emission, and partial oxidation of nitrogen and de-nitrification in anammox reactors. On the other hand, for hydrolysis reactions, the acceleration of cellulose hydrolysis, and compensation of optimum pH regions by potential shifting occurs according to Nernst's equations: $E0 = 0.17 + 0.06\ pH + 0.0074 \log pCH_4/pCO_2$.

Figure 4 shows a hypothetical principle of the potential controlled reductive metabolic control of reductive metabolism. The metabolic control of the supported enzymes or microorganisms on electro-conductive porous carbons can be changed by the bioreactor. The mediators, as electron shuttles, would control the metabolism by using conductive carbon materials with immobilized microorganisms. Figure 5 shows DGGE profiles of microbial analyses in the potential controlled bioreactor and hyper-thermophilic fermentation (65°C) for methane fermentation, compared with the microbial population prior to the processing. Although hyper-thermophilic fermentation markedly increased the Archaea population in association with the advance of methanogenesis in the reactors, the anaerobic fermentation in potential controlled bio-reactors increased the Archaea population slightly more than the hyper-thermophilic fermentation. Thus, anaerobic methane fermentation can be modified by the potential control.

Biogas produced through anaerobic fermentation of organic wastes in the biogas plants is a mixture of CH_4, CO_2 and other impurities. The combustible energy of CH_4 amounts to 890.8 kJ mol-1 (Haynes, 2016). As the methane content of biogas is approximately 50–70%, its high calorific value can be used for electric power generation. Additionally, hydrogen reformed from CH_4 can be harnessed to fuel cell as an alternative electric power co-generation system (CHP). However, biogas is often contaminated with hydrogen sulfide (H_2S) at high concentration. It is a corrosive compound for both the co-generator and the fuel cell. Fuel cells especially require an efficient biogas purification process. However, H_2S concentration in the biogas can vary between 1,000 and 20,000 ppm. At least, the H_2S concentrations for CHP must be reduced to less than 400 or 500 ppm. Bio-desulfurization is a fully microbiological process to eliminate H_2S using sulfide-oxidizing bacteria. Bio-desulfurization system will

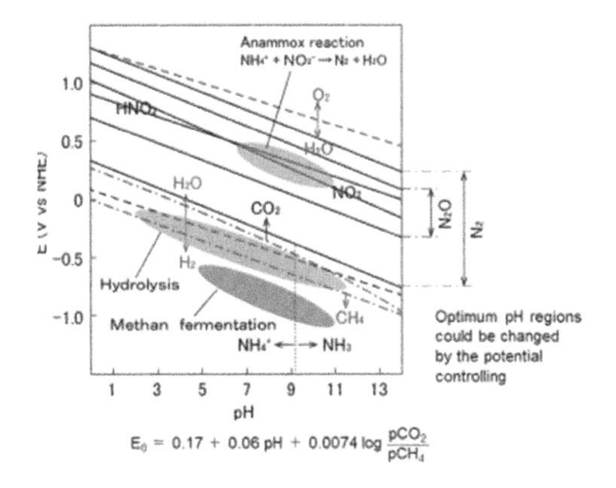

Figure 3. pH-potential diagram related to carbon and nitrogen.

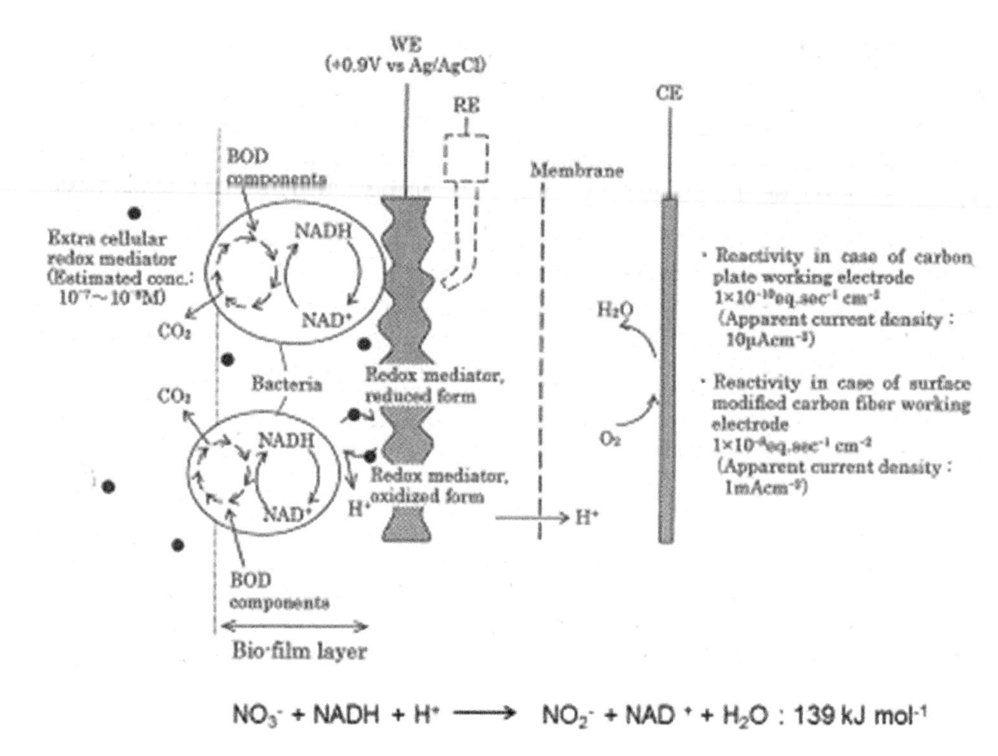

Figure 4. A hypothetic principle of the potential controlled reductive metabolism.

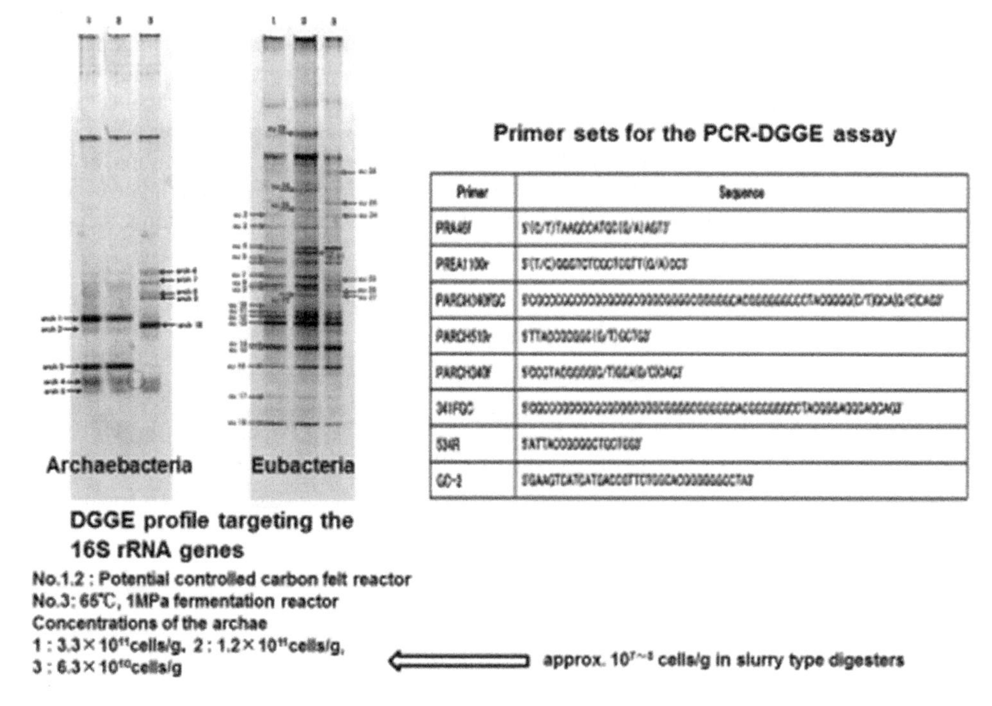

Figure 5. A microbial analyses of the potential controlled methane fermenters.

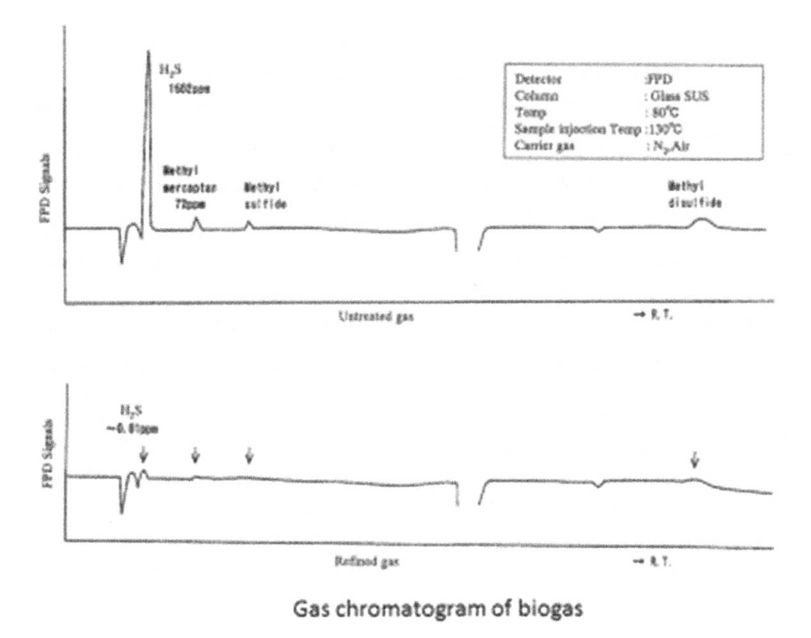

Figure 6. Bio-desulfurization of biogas with potential control.

effectively eliminate H_2S at low running cost without chemical reagents, unless unstable air supply occurs in the reactor. However, the microorganisms will become extinct if the air supply stops due to the power failure. Figure 6 demonstrates the process of bio-disulfurization ensured by the potential control. Even if the air supply to the microorganisms is unstable, bio-desulfrization will regularly work to eliminate H_2S through the control of redox state in the reactor.

4 CONCLUSION

Introducing potential controlled bioreactors can contribute to redox reactions, hydrolysis reactions and purification of biogas. Possible contribution to redox reactions comprises acceleration of methane fermentation, hydrogen fermentation selectivity, lactic acid fermentation selectivity, mitigation of nitrous oxide emission, as well as partial oxidation of nitrogen and de-nitrification in anammox reactors. On the other hand, for hydrolysis reactions acceleration of cellulose hydrolysis, and compensation of optimum pH regions by potential shifting occur according to Nernst's equations: $E0 = 0.17 + 0.06\ pH + 0.0074 \log pCH_4/pCO_2$. Moreover, introducing potential controlled bioreactors will contribute to the purification of biogas, i.e. bio-desulfurization of biogas ensured by potential control.

REFERENCES

Bendixen, H. J. 1997. Hygiene and sanitation requirements in Danish Biogas Plants. In: The Future of Biogas in Europe. J.B. Holm-Nielson, ed. BioPress, Risskov, Denmark pp.50–58.

Cel, W., Czechowska-Kosacka, A., Zhang, T. 2016. Sustainable Mitigation of Greenhouse Gases Emissions. *Problemy Ekorozwoju/Problems of Sustainable Development*, 11(1): 173–176.

Dai, X., Chen, S., Xue, Y., Dai, L.D., Li, N., Takahashi, J., Zhao, W. 2015. Hygienic treatment and energy recovery of dead animals by high solid co-digestion with vinasse under mesophilic condition: feasibility study. *J. Haz. Mat.* 297: 320–328.

Haynes, W. H., 2016, CRC Handbook of Chemistry and Physics. 97th ed. pp.5–68.
Kebreab, E., Clarek, K., Wagner-Riddle, C., France, J. 2006. Methane and Nitrous Oxide Emissions from Canadian Animal Agriculture: A review. *Can. J. Anim. Sci.* 86: 135–158.
Mao, C., Feng, Y., Wang, X., Ren, G., 2015. Review on research achievements of biogas from anaerobic digestion. *Renew. Sustain. Energy Rev.,* 45: 540–555.
Moset, V., Poulsen, M., Wahid, R., Højberg, O., Møller, H.B. 2015. Mesophilic versus thermophilic anaerobic digestion of cattle manure: methane productivity and microbial ecology. *Micro. Biotech.*, 8: 787–800.
Puckett, L.J., Timothy, K.C., Lorenz, D.L., Stoner, J.D. 1999. Estimation of Nitrate Contamination of an Agro-Ecosystem Outwash Aquifer Using a Nitrogen Mass-Balance Budget. *J. Environ. Qual.*, 28: 2015–2025.
Sasse, L. 1988. Biogas plants. Deutches Zentrum für Entwicklungstechnologien, GTH, Eschborn, Germany, pp.1–66.
Takahashi, J. 2010. Advanced Biogas Systems for a Renewable Energy Source. Renewable Energy 2010, Proceedings of Renewable Energy. O-Bm-10–2. pp.1–4.
Żukowska, G., Myszura, M., Baran, S., Wesołowska, S., Pawłowska, M., Dobrowolski, Ł. 2016. Agriculture vs Alleviating the Climate Change. *Problemy Ekorozwoju/Problems of Sustainable Development*, 11(2): 67–74.

HVAC systems supported by renewable energy sources—studies carried out at the Białystok University of Technology

M. Żukowski
Faculty of Civil Engineering and Environmental Engineering, Bialystok University of Technology, Bialystok, Poland

ABSTRACT: The paper presents the practical results of Regional Operational Program of Podlaskie Voivodeship implemented at the Bialystok University of Technology. The main objectives of the project were assessing the effectiveness of active and passive methods to improve the energy efficiency of infrastructure supported by renewable energy sources (RES) and improving the energy efficiency of Bialystok University of Technology (BUT) infrastructure through the use of RES. The results of the research work carried out by BUT Faculties are presented in the article. The author's projects related to the testing of renewable energy sources are also characterized. The last chapter of the paper includes research tasks that are planned to be completed in the future.

1 INTRODUCTION

"Climate and energy package" is a set of rules (2009/28/WE, 2009/29/WE, 2009/31/WE, 2009/406/WE 2009) designed by member states of European Union to combat climate change. According to these directives, the greenhouse gas emissions should be reduced by 20%, and the share of energy produced from renewable sources should simultaneously be increased by the same percentage. In the case of Poland, these indicators will probably not be achieved. It is planned that the share of RES in final energy consumption will increase to 15% in 2020, and only to 20% in 2030 (PEP2030 2010). Therefore, only widespread use of renewable energy sources can contribute to the fulfillment of "Climate and Energy Package" goals. RES have both advantages and disadvantages. The benefits can include:

- Potentially endless and sustainable source of energy.
- Usually exert a minimal impact on the environment.
- Rise of employment and economic benefits.
- Increase of energy security at the local level.
- Sometimes RES can be characterized by certain drawbacks listed below:

The cost of renewable energy technologies can exceed the cost of traditional technologies based on fossil fuels.

- Manufacture and implementation of RES can be costly.
- Wind farms will spoil the countryside views and can generate excessive noise intensity.
- Biomass burning is a source of atmospheric pollutants.
- RES can generate problems of balance of different power sources and problems of reliability of supply.

In order to perform practical assessment of renewable energy sources, the scientists from the Bialystok University of Technology developed a project co-financed by the European Union. Assumptions of this undertaking are described in the next section.

2 DESCRIPTION OF REGIONAL OPERATIONAL PROGRAM ON THE ASSESSMENT RENEWABLE ENERGY SOURCES

The main goal of the Regional Operational Program of Podlaskie Voivodeship, implemented at the Bialystok University of Technology, was to assess the use of renewable energy sources in the climate conditions of north-eastern Poland. This project consisted of two tasks:

- Study of the effectiveness of active and passive methods to improve the energy efficiency of infrastructure supported by renewable energy sources (scientific-research part).
- Improving the energy efficiency of BUT infrastructure through the use of renewable energy sources (investment phase).

Three tasks of this program were coordinated and co-financed by the Faculties and two tasks were carried out by the Bialystok University of Technology directly.

2.1 *Characteristic of the tasks included in EU project coordinated by Bialystok University of Technology*

The aim of the first tasks, directed by Żukowski, was to compare the thermal characteristics of the different types of water solar collectors. The solar domestic hot water system (SDHW) is located on the roof of the Hotel for Research Assistants at Zwierzyniecka 4 on BUT campus in Bialystok (Poland).

The object of the research project is composed of two parallel-connected systems. The first one, shown in Figure 1, consists of 35 flat plate collectors (FPC) with a total area of 72 m^2. The second system includes 21 evacuated tube collectors (ETC) with a total area of more than 74 m^2, which is shown in Figure 2.

The domestic hot water (DHW) is stored in eight tanks with the capacity of 1 m^3 each. They are located in the basement of the hotel. Two solar loops are separated from the DHW circuit through the compact heat exchangers and are filled with a propylene glycol solution that prevents freezing. The hydraulic diagram of SDHW is shown in Figure 3.

SDHW is equipped with a system for monitoring the operating parameters, which consists of 17 heat meters, 4 electricity recorders, and 42 PT 500 platinum resistance temperature sensors. Variables are recorded with the time interval equal to 1 minute. The basic parameters of ambient air and solar radiation are captured by the weather station (time step equal to 2 seconds) located on the roof of the hotel near the solar collectors. Detailed information about the project are given in (Żukowski & Radzajewska 2016, Żukowski 2016–1, Żukowski 2016–2)

Figure 1. Flat plate solar collectors connected in series of 5 units in one section (*photo Żukowski M.*).

Figure 2. Vacuum tube solar collectors connected in series of 3 units in one section (*photo Żukowski M.*).

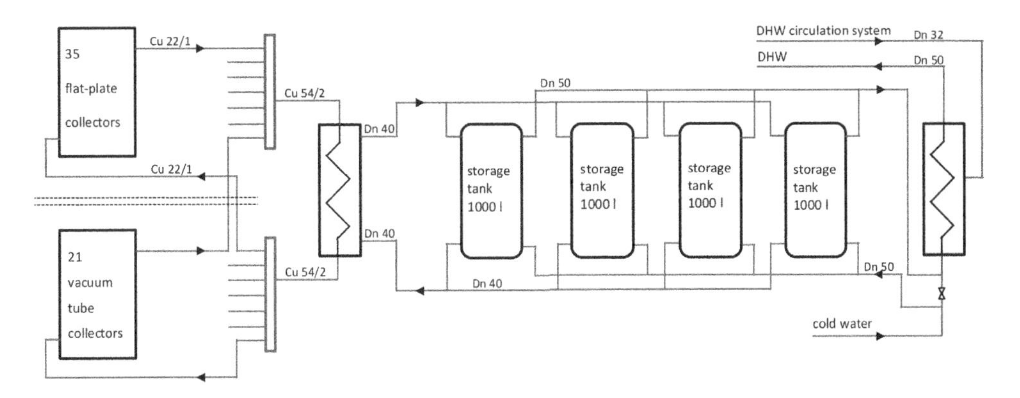

Figure 3. Hydraulic diagram of a DHW system supported by solar panels (*prepared by Żukowski M.*).

After two years of operation, the solar system provided 134036 KWh of thermal energy. Figure 4 shows the amount of energy produced by month during the entire study period. Currently, the difference in energy gains is around 2.9% in favour of flat-plate collectors that are two times cheaper than vacuum tube panels.

The analysis of experimental results shows that the vacuum tube collectors are characterized by higher power in winter. This is due to lesser heat loss from evacuated tubes in comparison to the flat panels. Unfortunately, at this time of the year the intensity of solar radiation is very low.

As shown by the results of this research, useful energy extracted from the collector (annual energy gain) related to the m^2 of total area was equal to 412.3 $kWh/m^2/a$ (FPC), 458.5 $kWh/m^2/a$ (ETC) in 2015, and 365.9 $kWh/m^2/a$ (FPC), and 389.3 $kWh/m^2/a$ (ETC) in 2016.

The aim of the second task, coordinated and co-financed by the Bialystok University of Technology, was to develop new technical solution and build a biogas plant operating in dry technology—manure, grass silage, straw. This project, called "Optimizing the production of agricultural biogas", was carried out under the supervision of prof. P. Banaszuk. This technology of biogas plant is suitable for use in a single agricultural farm. The advantage of this system is the full utilization of heat for farming purposes. The small biogas plant is located on A. Naumczuk farm in a short distance from Bialystok. Currently, scientists from Department of Environmental Protection and Management are working on the optimization of biogas production from this research installation.

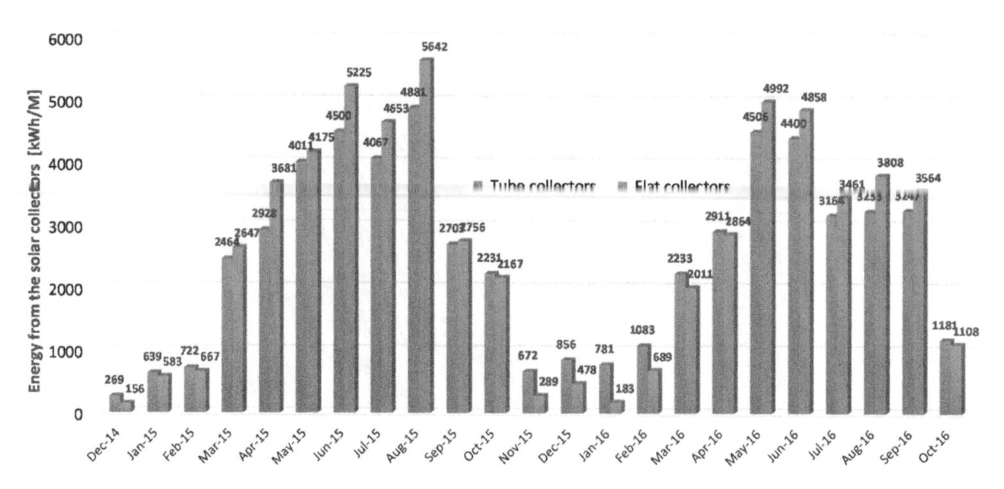

Figure 4. Monthly energy gain from solar domestic hot water system (*prepared by Żukowski M.*).

Figure 5. Compact biogas plant designed by the HVAC design office form Germany (*photo Banaszuk P.*).

2.2 *Characteristic of the task included in EU project coordinated by Faculty of Architecture*

Researchers of the Faculty of Architecture developed the design project and construction concept for energy-saving building. Dr. Turecki was the person responsible for creating a design and for coordinating this complex project. The building of research laboratory for testing renewable energy sources, shown in Figure 6, is located just behind the Faculty of Architecture. Technical solutions used in the building include:

- Meteorological station.
- Central heating system using a heat pump.

Figure 6. The research laboratory for testing renewable energy sources—view of the southern façade (*photo Żukowski M.*).

- Flat plate solar collector for DHW system located on the roof.
- Vacuum tube solar collector integrated with the vertical façade.
- Photovoltaic cells.
- Active wall façade.
- Small wind turbine with horizontal axis.
- Mechanical ventilation system with recuperation and integrated with the ground heat exchanger.
- Lighting automatic system.
- Thermal energy storage systems.
- Monitoring and control system (BMS) of parameters for each technology with computer visualization integrated with the Internet.

The dimensions of the building are: length – 13.05 m, width – 7.55 m, height – 9.03 m. The other technical parameters of the building are as follows: number of floors – 3, gable roof (slope of 38 degrees), usable area – 185.62 m^2, building area – 98.20 m^2, cubic volume – 1116.33 m^3.

2.3 *Characteristic of the tasks included in EU project coordinated by faculty of mechanical engineering*

The project coordinated by Faculty of Mechanical Engineering consisted of two phases:

- Optimization of vegetable oil properties as a fuel.
- The use of vegetable oil as fuel in the vehicles engines.

The scope of the first task involved:

- Buying oil mills (with a set of filters), testing the manufacturing process and purification of the oil.
- Examination of the effectiveness of different kinds of filters.
- Test possibilities for effective removal of phosphorus, calcium and magnesium from the oil.
- The construction of the necessary infrastructure related to the study process.

Figure 7. Engine braking with a complete Perkins tractor engine (*photo Żukowski M.*).

The second task was far more difficult and consisted of:

- Determination the maximum safe admixture of vegetable oil as a fuel component added to the oil used in the existing diesel engines.
- Adaptation of the diesel engines to burn pure vegetable oil using additives and without them (Figure 7).

2.4 *Characteristic of the task included in EU project coordinated by faculty of electrical engineering*

The scientific goal of this project was to carry out comprehensive research of the complementary hybrid system for electricity generation based on solar and wind energy coupled with the development of technical and economic analyses. The hybrid system (Figure 8) is based on the use of energy generated by photovoltaic panels and wind generators. It was assumed that the optimal system for supplying the reference building should have the capacity of 20 kWp: 10 kWp supplied from wind turbines, and 10 kWp supplied from PV panels. A hybrid power station can operate implementing two alternative options: supply power to the power system of BUT campus, as well as provide power to the electricity network. This task was developed under the direction of dr. W. Trzasko.

Photovoltaic system is located on the roof of the building at Zwierzyniecka 10 on BUT campus and consists of:

- 12 PV units with the total area of 19.48 m^2 and angle of inclination of approx. 38 degrees to the horizontal, southern direction (DC 3.0 kWp).
- 12 PV units with the total area of 19.48 m^2 with servo system (tracker) that keeps track of biaxial movement of the Sun across the horizon by the astronomical clock (DC 3.0 kWp).
- 6 PV units with the total area of 9.74 m^2 located on south—eastern façade of the building—angle of inclination equal 90 degrees to the horizontal (DC 1.5 kWp).
- 6 PV units with the total area of 9.74 m^2 located on south—western façade of the building—angle of inclination equal 90 degrees to the horizontal (DC 1.5 kWp).

The project also included the installation of two wind turbines:

- The Darrieus wind turbine with vertical axis PowerWind 5000, tri-wing, diameter – 3.5 m, height of blades – 3 m, mast with a height of 15.61 m (DC 5 kWp).

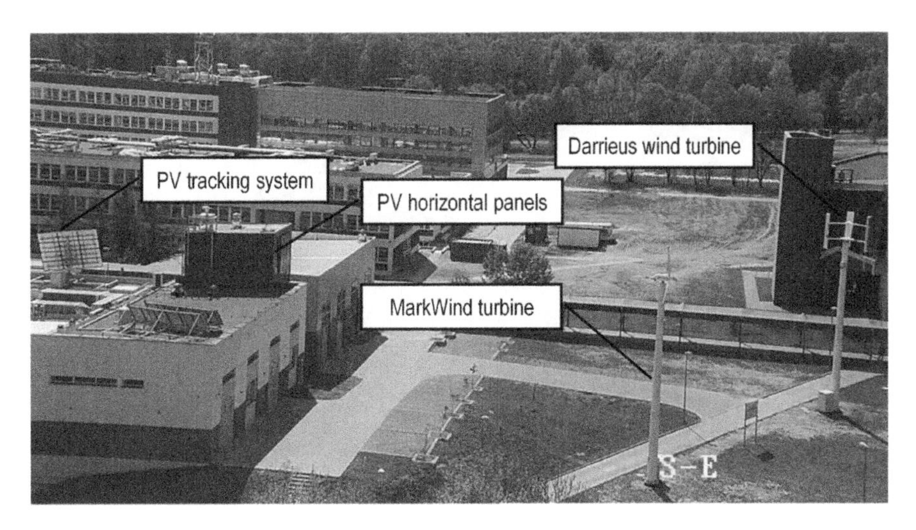

Figure 8. General view of the hybrid power plant (*photo Żukowski M.*).

- Wind turbine with horizontal axis MarkWind 5000P, rotor diameter – 4.8 m, mast with a height of 15.27 m (DC 5 kWp).

Data analysis showed that the total production of electricity from February 1, 2015 to December 3, 2016 was 18995 kWh, including: PV cells – 18083 kWh, and wind turbines—only 912 kWh.

2.5 *Characteristic of the tasks included in EU project coordinated by faculty of civil engineering and environmental engineering*

Faculty of Civil and Environmental Engineering implemented the project on retrofit thermal insulation (Święcicki et al. 2014–1, 2014–2) and modernization of the central heating system, using renewable energy sources (Krawczyk & Gładyszewska 2015). Retrofitting included elimination of thermal bridges, as well as insulation of external walls (Figure 9) and roofs. Before the retrofit works began, the index of the annual demand for final energy EF of Buildings A and B of Faculty of Civil and Environmental Engineering was 270 kWh/m^{2}a. After the implementation of all modernization projects, the EF index decreased to 105 kWh/m^{2}a (Święcicki et al. 2014–1, 2014–2).

Modernization of the boiler room included the use of brine-to-water heat pumps (2 units with a capacity of 117.2 kW each and 2 units with a capacity of 21.2 kW each) as a heat source for all central heating systems in three buildings of the Faculty of Civil and Environmental Engineering.

The heat from the ground is provided by vertical borehole heat exchangers: 52 vertical loops with the of depth 100 m each, 10 loops with the depth 50 m at an angle of 45 degrees, and 10 loops with the depth of 50 m at an angle of 55, 60, and 65 degrees. Operation of heat pumps and geothermal heat exchangers is monitored using heat meters and several hundred temperature sensors (Piotrowska & Woroniak 2015).

Further devices using renewable energy sources were used in the modernization of the ventilation system (Żukowski 2012). Two ground air heat exchangers (GAHE) and air ventilation units with the heat recovery were installed. The purpose of this research project was a comparative analysis of two types of GAHEs under the same operating conditions. Both exchangers occupy the same ground surface (length – 28 m, width – 18 m).

First unit—multiple-pipe earth-to-air heat exchanger, shown in Figure 10, is a new design that has not been previously installed in Poland. Figure 11 shows the next device—the

Figure 9. View of Building B of Faculty of Civil and Environmental Engineering during assembly of the thermal insulation (*photo Żukowski M.*).

Figure 10. Multiple-pipe GAHE during assembly (*photo Żukowski M.*).

Figure 11. Flat-plate GAHE during assembly (*photo Żukowski M.*).

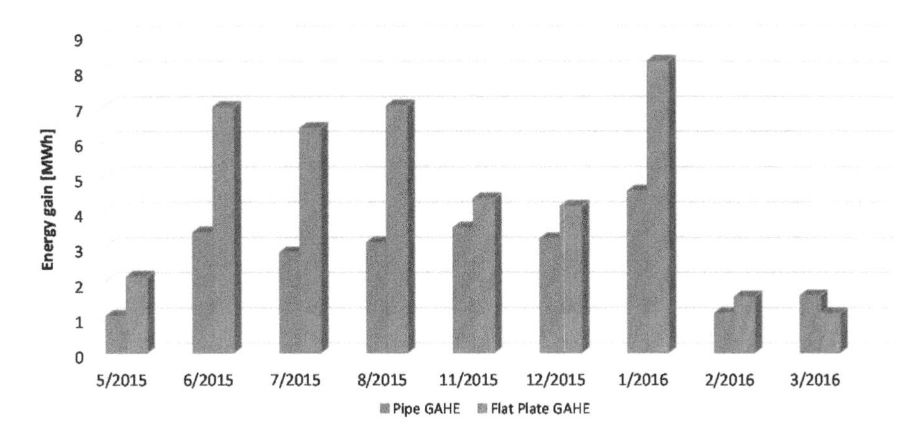

Figure 12. Energy gains coming from GAHEs (*prepared by Żukowski M. & Topolańska J.*).

flat earth-to-air heat exchanger. An advantage of this configuration is that the air flowing through this GAHE is in direct contact with the ground.

As shown by the results of experimental tests (Figure 12) the flat-plate GAHE is characterized by better energy efficiency. Its thermal and cooling loads are higher both during the winter and summer. Additionally, the air flowing through this device is moistened by contact with the ground.

3 PRESENTATION OF OTHER RESEARCH PROJECTS ON EXAMINATION OF RENEWABLE ENERGY SOURCES

Numerous studies and analyses of systems using renewable energy sources are conducted in the Department of HVAC Engineering of BUT. The first example is the project on experimental testing of ceramic solar collectors (Figure 13) not performed in Europe up to now. Tests of these devices were only carried out in China, albeit limited in scope. A special test rig, shown in Figure 14, has been made for this purpose. It can be used to determine the thermal performance of each type of liquid solar collectors.

Preliminary studies have been carried out in summer 2016. The efficiency is the most important parameter of the thermal performance of solar panels. As it turned out, the ceramic solar collector zero loss efficiency is about 0.65 (see the graph in Figure 15), which should be classified as a good result.

Thermal characteristic of the newly designed solar air heater (SAH) for preheating ventilation air was tested in the summer of 2016. The new type of SAH (Figure 16) is equipped with turbulators (not yet used in such devices) for the intensification of the heat exchange. Testing the new type of solar panel was the joint project of BLDC—SOLAR Company and Department of HVAC Engineering.

The efficiency of the tested SAH was in the range of 50%–70% (see the graph in Figure 17). The energy efficiency of the tested system increases with the intensity of solar radiation. Thus, it can be concluded that the tabulators had a positive effect on increasing the energy gains.

Another way to increase the exchange of heat inside the solar air heater is the use of micro-jet.

Experimental investigation of a new construction of SAH was performed within the framework of a Grant No. N N523 615539 from the Polish Ministry of Science and Higher Education "Experimental and modelling study of a heat exchanger based on micro jets impinging on a heat exchange surface". The general principles of the design and operation of the solar panel is presented in Figure 18. A detailed report of the research is in the papers (Żukowski 2013, 2015). The invention was patented in Patent Office of the Republic of Poland (PL 221962 B1; WUP 06/16).

Figure 13. Flat-plate ceramic solar collectors (*photo Żukowski M.*).

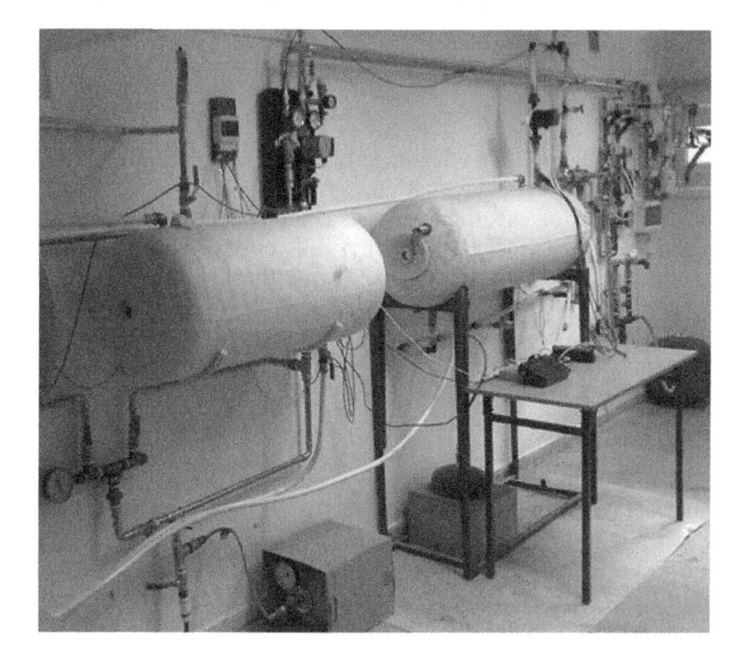

Figure 14. Experimental setup for testing liquid solar panels (*photo Żukowski M.*).

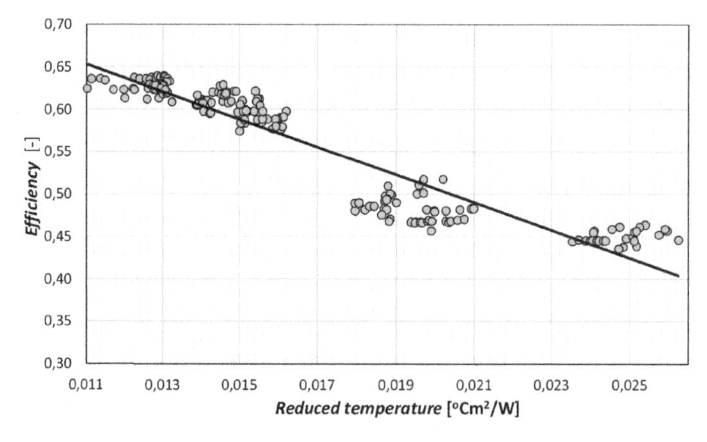

Figure 15. Efficiency of the ceramic solar collector as a function of reduced temperature difference (*prepared by Żukowski M. & Woroniak G.*).

Figure 16. SAH tested experimentally in normal operating conditions (*photo Żukowski M.*).

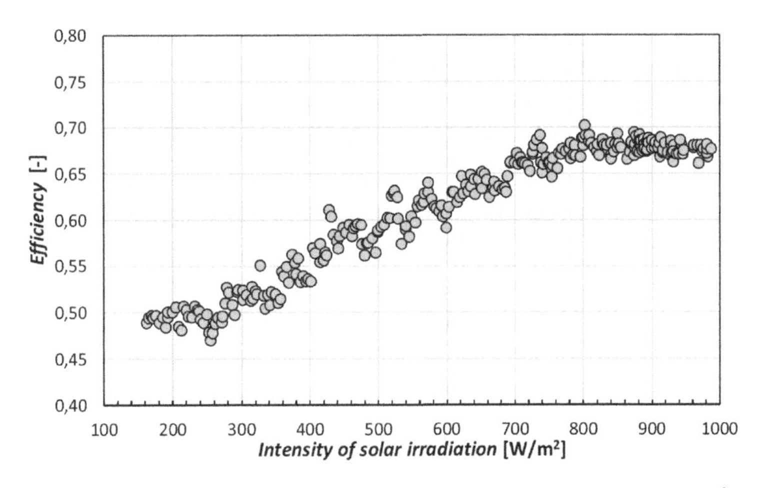

Figure 17. Thermal efficiency of SAH as a function of solar radiation (*prepared by Żukowski M. & Woroniak G.*).

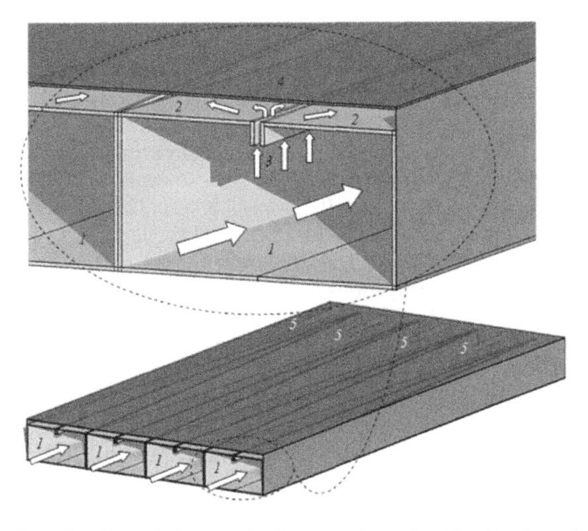

Figure 18. Scheme of a microjets air heater: 1 – bottom duct, 2 – distribution plate, 3 – slot jet nozzle, 4 – absorber plate, 5 – return channel (Żukowski 2015).

4 SUMMARY AND FUTURE WORK

The possibility of applying renewable energy sources in the climate conditions of north-east of Poland was tested in the Regional Operational Program of Podlaskie Voivodeship. The project was completed at the end of 2015. Its result was the creation of many objects and test benches. Currently, the research teams from BUT continue the previously started research projects. Future work of the author that will focus on renewable energy sources and include the tasks listed below.

- The analysis of the energy balance of Municipal Waste Treatment Plant in Bialystok.
- Continuation of experimental testing of ceramic solar collectors including determination of thermal characteristics in transient operating conditions based on ISO 9806 (2013).
- Continuation of experimental testing of two systems of solar collectors on the roof of Hotel for Research Assistants including determination of thermal characteristics each group of collectors connected in series on the basis of the Hottel—Whillier—Bliss (HWB) model.
- Continuation of experimental testing of GAHEs including determination the energy gains in winter and during the summer in different operating conditions.
- Development of virtual laboratories (e-labs) of RES within project financed by the European Union—VIPSKILLS.

ACKNOWLEDGMENTS

This work was performed within the framework of a Grant of Bialystok University of Technology (Grant No. S/WBIIS/4/2014).

REFERENCES

ISO 9806:2013 Solar energy—Solar thermal collectors—Test methods.

Krawczyk, D.A. & Gładyszewska-Fiedoruk, K. 2015. Changes in energy consumption, greenhouse gasses emission and microclimate in classes after thermomodernization, 5th International Conference on Environmental Pollution and Remediation 15–17.07, 2015 Barcelona.

PEP2030 (2010). Polish Energy Policy until 2030. Annex to Resolution No. 157/2010 of the Council of Ministers.

Piotrowska, J. B. & Woroniak, G. 2015. Preliminary results of the temperature distribution measurements in the soil around the vertical ground heat exchanger tubes, 2nd International Conference Renewable Energy Sources engineering, technology, innovations, May 26–29, 2015 Krynica, Poland.

Święcicki, A., Sadowska, B. & Sarosiek, W. 2014–1. Termomodernizacja budynku WBiŚ. Cz. 1. Stan istniejący na podstawie dokumentacji archiwalnej i pomiarów. *Rynek instalacyjny*, 10: 36–38.

Święcicki, A., Sadowska, B. & Sarosiek, W. 2014–2. Kompleksowa termomodernizacja budynku WBiIŚ Cz. 2. Plan inwestycji z analizą potencjału efektów termomodernizacji. *Rynek instalacyjny*, 11: 21–25.

Żukowski, M. 2012. Ocena potencjalnych możliwości zastosowania GPWC na terenie Politechniki Białostockiej. *Ciepłownictwo, Ogrzewnictwo, Wentylacja*, 8: 323–326.

Zukowski, M. 2013. Heat transfer performance of a confined single slot jet of air impinging on a flat surface. *International Journal of Heat and Mass Transfer*, 57: 484–490.

Zukowski, M. 2015. Experimental investigations of thermal and flow characteristics of a novel microjet air solar heater. *Applied Energy*, 142: 10–20.

Żukowski, M., Radzajewska P. 2016. Optymalny rozstaw kolektorów słonecznych. *Ciepłownictwo, Ogrzewnictwo, Wentylacja*, 1: 8–11.

Żukowski, M. 2016–1. Analiza zagadnień związanych z zaleganiem śniegu i oszronieniem kolektorów słonecznych. *Ciepłownictwo, Ogrzewnictwo, Wentylacja*, 5: 171–175.

Żukowski M., 2016–2. Assessment of the environmental effects resulting from the use of solar collectors. *Rocznik Ochrona Środowiska*, 18(2): 284–293.

Status and trends of PV agriculture in China

M. Zhao, P. Xu & X. Duan
Nanjing Agricultural University, Nanjing, China

Z. Cao
Institute of Soil Science Chinese Academy of Sciences, Nanjing, China

L. Pawłowski
Lublin University of Technology, Lublin, Poland

ABSTRACT: Photovoltaic agriculture (PV) is a new type of agricultural production which combines solar power generation and modern agriculture. It maximizes land utilization and realizes "one place and dual use"; meanwhile, plants grow better under its shade. The research analyzes the present situation of PV agriculture in China, presents the four typical development models and includes a benefits analysis for the four typical models by the method of document retrieval, data analysis, statistical investigation and case analysis. Finally, the paper puts forward the development trend of PV agriculture, the research emphasis and suggestion for promoting the development of PV agriculture.

Keywords: PV agriculture; status quo; models; benefit; trend; suggestion

1 INTRODUCTION

PV agriculture is a new type of agricultural production, which combines solar power generation and modern agriculture. Although solar power is widely used in modern agricultural research, plantation, breeding, irrigation, pest control and agricultural machinery as power supply (Liu 2014, Liu 2015, Żukowska et al. 2016, Żelazna and Gołębiowska 2015, Chakraborty et al. 2016, Bielińska et al. 2015). PV agriculture is a complex system of both photoelectric and agricultural production which can be conducive to agricultural development and improve the efficiency as well as simultaneously produce solar power. PV agriculture originated in the concept of agricultural photovoltaic which was proposed by the Japanese mechanical engineer Akira Nakajima (Jiang et al. 2016), according to the definition of "point of light saturation", combining the solar panel system with agricultural production. At present, China's PV agriculture is in the early stage of development, immature in the mode, application etc., and thus it lacks systematic and theoretical research. The research on PV agriculture in Japan and Europe pertains to the combination of photovoltaic modules and greenhouses, the type of photovoltaic components and power generation efficiency (Yang et al. 2015).

As a new model of agricultural development, PV agriculture has an obvious effect on improving the land utilization, realizing cleaner production and increasing the value of agricultural output. In recent years, the construction of ecological agriculture and ecological civilization which were emphasized and advocated, by the China's government, resulted in PV agriculture ushering in an unprecedented development opportunities. This research reviews the status of PV agriculture development, trends and difficulties presented in literature, statistical investigation, data collection and typical case analysis, so as to provide a reference for the development of PV agriculture.

2 THE PRESENT STATUS OF PV AGRICULTURE

2.1 *State of development*

2.1.1 *PV industry has laid a good foundation for the development of PV agriculture*

PV agriculture has well promotion greatly based on the development of PV industry. The development of Chinese PV industry is already one of the fastest around the world. In recent years, the total installed capacity of PV generation showed a linear upward trend. In 2015, the installed capacity of PV power generation in China was 43 GW, making it the country with the largest installed PV power capacity. According to the latest data from China Energy Bureau, the total installed capacity reached 50.31 GW in the first quarter of 2016, which constitutes a 52% increase in relation to the same period of previous year (National Energy Board of China, 2016). The "13th Five-Year-Plan" (2016–2020) also proposed a clear goal for the future development of PV power capacity: by 2020, China is to achieve the PV installed capacity of 150 GW. As can be seen from the above-mentioned data, the Chinese PV market has a great potential for development.

2.1.2 *Rapid and regional development of China PV agriculture*

Distributed PV power generator is mainly installed in the eastern China, and PV agriculture is an important component of it. In 2013, the installed capacity of distributed PV power was 3.1 GW. in which, the approved agricultural-light complementary and fishing-light complementary projects totaled 2.9 GW, accounting for 94% of the distributed PV power. In the first half of 2014, more than 2.0 GW of PV agriculture projects are under construction or have been contracted (Fang et al. 2015). The installed capacity of PV agricultural projects will reach 6 GW in 2015 according to the installed capacity of distributed PV.

The eastern region, including coastal provinces, such as Zhejiang, Jiangsu, Shandong, Guangdong etc., is highly developed in regard to modern agriculture, involving agricultural facilities, fishery farming and so on, as well as a large number of PV power generation systems. Due to the limitation of the per capita land, agriculture, and fishing are popular branches. For example, more than 2 million kilowatts of PV agriculture are expected to be achieved in Zhejiang Province by the end of 2016. The development of PV agriculture in the eastern and western China is characterized by an obvious distinction; the eastern region prefers PV facilities connected to agriculture and fishery, while the western region prefers PV related to livestock. In Inner Mongolia and its adjacent regions, PV livestock farms are being developed, in addition to roof-mounted PV panels and PV panels installed on pastures.

2.1.3 *Diversified PV agricultural projects and integrated development*

PV agriculture has been developed in different parts of the country, and more and more agricultural projects are beginning to be combined with PV. Xiang-gen Gao mentioned that PV

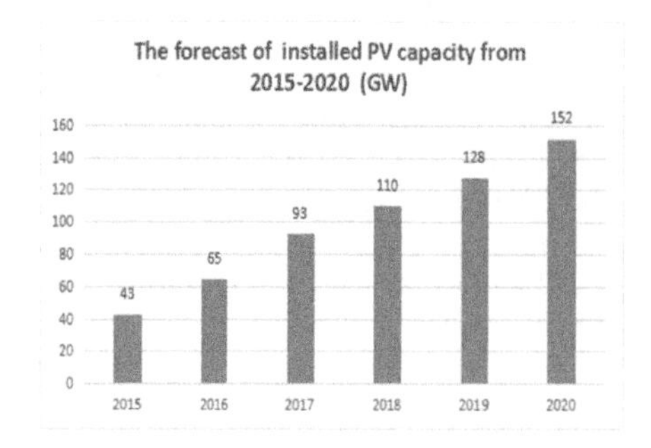

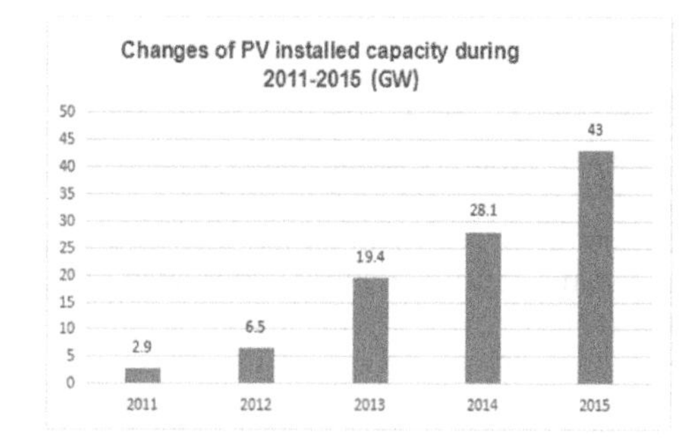

Figure 1.

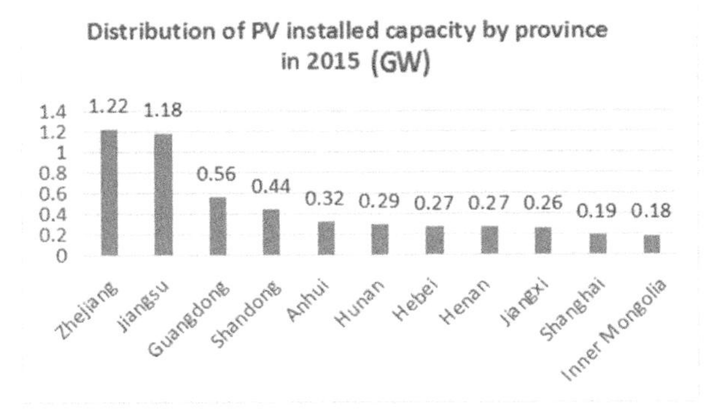

Figure 2.

agriculture now has ten modules, related to: vegetables, livestock and poultry, edible fungi, medicinal plants, bushes, ecological and farmhouse, thermoelectric, water, and service. This paper mainly reviews four typical PV agricultural models, including PV-horticulture, PV-livestock/poultry, PV-fishery and PV-tea. A typical model has many sub-modules, for example, PV horticulture consists of PV-edible fungi, PV-herbs, PV-vegetable and fruit, etc.; this makes PV-agriculture highly diversified.

PV-agriculture can also involve agricultural facilities, tourism, and PV power generation set in one place. According to the local conditions, the model may be combine the functions of production, tourism, science and education etc, showing the development of integration, aiming to maximize efficiency.

2.1.4 *Standardized development of PV agriculture*

The China PV Agriculture Standards Seminar held in 2016, published the first batch of PV agricultural groups and enterprise standards in China. These pioneered the world's standard of PV agriculture, including: "Technical Guidelines for PV Agricultural Systems, PV Terms and Symbols", "fishery light pond and the park construction, aquaculture management and technical specifications", "photovoltaic kiwi fruit orchard technical guidelines", "PV oil and peony garden technology guidelines", "photovoltaic power plant system design, construction, inspection and maintenance norms". The introduction of group standards will make up for the long-term lack of standards, assuming a normative role for PV agriculture. However, these standards should be further improved, in order to meet national/international requirements.

2.2 *Development bottlenecks*

2.2.1 *The innovation of technology integration needs a breakthrough*

The main difficulty of integrated innovation of PV agriculture lies in the need for a technological breakthrough. Agricultural production is seriously affected by the weather, region, land, environment and other external conditions. Therefore, the constraining factors will increase after the integration of PV generation and agriculture. In order to ensure the better integration, taking into account the variety, design, management, profit model and other factors is essential, but the current technological maturity needs to be improved.

2.2.2 *Lack of specialists and professional researcher and institutions*

PV agriculture is the fusion of PV and agriculture. In China, there is no systematic theoretical research. Due to the interdisciplinary nature of PV agriculture, it is necessary to train interdisciplinary specialists who are competent at both PV and agriculture. In addition, there are no institutions and no special subject system in colleges, resulting in a lack of specialists and practical research achievements in this area.

2.2.3 *Basic theory and applied technology research needs to be deepened*

PV agriculture is involved in the field of both PV power and agriculture. Both PV power and agriculture alone have formed a theoretical system, but these two components still lacks of a deep and systematic study. For example, there is no study pertaining to the best combination of PV and agriculture to achieve higher light use efficiency, and for the impact of PV agriculture on the ecology and environment, etc.

2.2.4 *Standard system need to be improved*

PV agriculture is in the initial stage of development and lacks uniform technical management standards. Despite the rapid development of PV agriculture projects around the country, the design and scale of the PV projects also lack professional standards (Hanergy Holding Group Limited, 2015). This lack of standards may easily lead to chaos. PV agricultural standard system should focus on the technical, design, size and access standards based on the basis of this practice, which should be constantly amended.

3 THE DEVELOPMENT MODEL OF PV AGRICULTURE

Due to the lack of in-depth theoretical research, the models of China's PV agricultural, which are in the early stages of development, remain uncertain. According to the industry classification, one can enumerate PV Horticulture, PV Farming, PV Forestry, etc. On the other hand, the solar energy application, distinguishes between PV related to water conservancy, cottage, greenhouses, solar insecticidal lights, solar water purification systems etc. (Zhang et al., 2015).

This study mainly reviews four typical PV agriculture modes: PV Horticulture, PV Livestock and Poultry, PV Fishery and PV Tea (bushes), which provide higher economic benefit and better development prospects in China.

3.1 *PV horticulture*

Broadly speaking, horticulture refers to the techniques of cultivation and breeding vegetables, fruit trees, flowers, Chinese herbal medicines, edible mushrooms, and ornamental seedlings. PV horticulture is a combination of photovoltaic power generation and horticulture production, a dual-use of the land. In involves the choice of specific cultivation, breeding species, as well as general consideration to their growth habits, economic efficiency and other factors. From the point of view of economic benefits, the sub-models such as PV fungi, PV vegetables, PV Chinese herbal medicine, PV flower seedlings and PV fruits are currently most popular in China.

3.1.1 *PV-edible fungi*

This is the most proper model, complementary between the edible fungi and light. At present, this model has a stable promotion and operation. It contributed to the success of Qingdao Huasheng Green Energy and Liaoning Fuxin.

The species

The normal growth of edible fungus products requires appropriate temperature and humidity, well-ventilated, scattered light environment (Wang, 2012). Various strains of edible fungi require different light intensity and light quality, with suitable temperature and humidity. Common edible fungi, such as mushrooms, black fungi, tremella, nyingchi etc., are characterized by a good taste and health value.

Construction characteristics

The structure is generally built through the establishment of a PV shed. The shed design commonly uses steel structure; the solar panels are installed on the roof. Mushrooms have very low light requirement but need a relatively high temperature. Therefore, the whole shelter can be built in single-family greenhouses.

Complementary features between light and agriculture
The patterns are generally achieved by establishing a PV shed. Common PV sheds have a multi-span glass or film greenhouses, and/or single-family glass or film greenhouse. The design of a bacterial shed uses steel structure of a roof covered with solar panels. This guaranteed the adequate amount of light for power generation and the growth of edible fungi.

3.1.2 *PV-fruit and vegetables*

The complementary model of light and fruit and/or vegetable mainly refers to another application of PV horticulture: shed power generation and cultivation of fruits or vegetables in a greenhouse shed. Based on the conditions in the north and the south of China, the structure of a PV greenhouse will be different. In general, one can choose the varieties characterized by higher yield and quality, offering better economic returns, but the light demand of these crops should also be considered.

The species
In the choice of vegetables and/or fruit varieties, one must consider the crop growth conditions and the market conditions of the product. For instance, tomato, okra, etc. can be chosen in PV greenhouse; alternatively, kiwi fruit, strawberries and other fruits may be cultivated for higher economic returns.

Construction characteristics
In the northern region of China, the PV panels of a greenhouse are installed directly on the roof; for the southern region of China, one can choose an open multi-span greenhouse for cultivation of various crops all year round.

Complementary features between light and agriculture
Installing the solar PV panels on the roof will affect the absorption of light by crops. Hence, it is necessary to extend the time of the plant light and enhance the greenhouse temperature in an alternative way (Gou, 2014).

3.1.3 *Flower seedlings*

The species
Varieties of flowers characterized by a high economic value and relatively high shade-resistance are the main choice. These include Cymbidium, Phalaenopsis, Anthurium, Alpine azaleas, etc., as well as meaty plants or colorful ornamental seedlings.

Construction characteristics
Usually, linked greenhouses with roofs covered by PV panels are chosen.

Complementary features between light and agriculture
Selection of semi-shading greenhouses can reduce the amount of direct sunlight available for plants and the top of the PV panels can make full use of solar energy conversion.

3.1.4 *PV-Chinese medicinal herbs*

The species
Chinese herbal medical plants with traditional Chinese medical characteristics, constitute an important supplement products in aiding medical treatments and promote one's health. Usually, the varieties of herbs which are shade tolerant and have a higher economic value are chosen. These include Ganoderma lucidum, and common species such as Peony, Sophora flavescens, Astragalus, Radix, Dendrobium etc.

Construction features
The PV Chinese medical herbs cultivation can be divided into outdoor and indoor planting. In the outdoor planting, the PV panels are placed in the ground, whereas the indoor planting mainly involves the construction of PV greenhouses. At present, this model is widely promoted.

Complementary features between light and agriculture
Setting up PV panels can provide a low light environment for the growth of Chinese herbal medical plants. It is conducive to the growth of these herbs and achieves double benefits of power generation and the growth of medical plants.

3.2 *PV livestock breeding*

PV animal husbandry includes livestock breeding (cattle, sheep, pigs, etc.) and poultry breeding (chickens, ducks, etc.). By installing PV panels in pens or breeding greenhouses, the model can maximize the use of land resources and achieve full coverage of the shed roof. The first Chinese PV pig farm was completed in Hebei Dingzhou, the electricity is supplied by the PV panels installed on the roof (Yiming, 2013).

The species
PV livestock breeding includes poultry farming and animal husbandry. Poultry species are mainly chicken, duck etc., and animal species are mainly pig, cattle, sheep, etc., choosing the varieties of high economic value on the market is one of the key points for successful business.

Construction features
Installing PV panels on the roofs has no effect on the growth of breeding objects. The requirements for the arrangement of components, inclination and other aspects are low. The layout of PV panels on breeding greenhouses will affect the light transmittance, so panels should be installed in the shed behind the roof in order to ensure an adequate light transmission rate.

Complementary features between light and agriculture
Installing PV panels on the roof has no effect on livestock and poultry growth. However, laying PV panels on greenhouses will affect the light transmittance, temperature, and the growth rate in the object. Lights, heat lamps and other related measures are required to solve this problem, to ensure better environment for good growth of livestock and poultry.

3.3 *PV fishery*

In PV fishery, PV panels are installed above the water surface to achieve a parallel development of aqua-cultured fish/shrimp breeding and power generation (Zhao et al., 2013). East China and South China regions have a wide distribution of water bodies which are suitable for the popularization and application of the complementary patterns of fishing and solar energy generation. With the solar power, this model can effectively control the water temperature and pH, aerobic conditions, electric water sterilization, as well as automatic feeding, and other modern fishing facilities, in order to maintain a good breeding environment and improve the quality of aquatic products.

The species
Aquaculture mainly involves fish and crustacean species with high market value, such as herring, crab, eel, snail, loach and so on.

Construction features
Solar PV panels installed above surface of water include closed, open, floating and leaping types.

Complementary features between light and agriculture
In general, PV panels installed over a large area will significantly reduce the light transmission rate affecting the growth of certain species of fish but has a lesser impact on other. This model can take full advantage of the space above the water surface, but can also ensure that the fishery part operates while the space is fully utilized, thus leading to a win-win situation.

3.4 *PV tea*

Covers of PV panels slightly obscure the sunlight (scattered light) shining on the tea bushes, but the strong shade of tea trees provides a mild environment for tea growth and it greatly improves the quality of tea leaves regarding the content of aminophenol. Intelligent installation of the shed system and usage of solar energy power enables remote monitoring, automatic temperature and humidity control, fertilization, watering, etc. for a tea garden. Currently, this new model has been widely used in Shandong, Hubei, Jiangsu, and other places.

The species
Tea trees growing in warm climates do not tolerate cold, excessive sunlight or the environment which is too alkaline or too wet (Huasheng, 2015). Chinese tea mainly includes six kinds: green tea, black tea (red tea), white tea, yellow tea, wulong tea and poole tea, divided according to post harvest processing techniques.

Construction characteristics
Generally, tea trees are planted under open-type PV greenhouses. PV panels are only installed on the roof of open-type PV greenhouses. The temperature, humidity, light transmittance, ventilation, and density of tea trees are adjusted to create the optimum conditions for tea growth.

Complementary features between light and agriculture
PV greenhouses can block the strong sunlight for the tea trees, provide a mild environment for the tea growth, and improve the content of tea aminophenols to ensure the high quality of tea without affecting the growth of tea under the roof, thus fully utilizing the space to achieve power generation income.

4 THE BENEFITS ANALYSIS OF PV AGRICULTURE

4.1 *Economic benefits*

The development of PV agriculture can maximize the land utilization rate and create a win-win situation for agricultural bio-economy and solar power generation. The economic benefit of different PV agricultural models has been analyzed.

4.1.1 *Horticulture*
For example, in order to cultivate edible fungi, a multi-span film greenhouse was used for cultivation of Pleurotus eryngii. The effective cultivation area of 1 hm 2, greenhouses are covered with PV panels, while the total installed capacity of PV power generation is 600 KW.

Table 1. Economic analysis of PV-edible fungi.

	Facilities & input	Total Investment (10^4 yuan)	Annual income (10^4 yuan)
Agricultural Facilities	Multi-span film greenhouse	200	From edible fungi: 30
	Land rent	1.5	
	Rods and breeding materials	3	
	Management and maintenance	1.2	
PV Facilities	Photovoltaic facilities	480	From generated energy: 114
	Total	685.7	144

As can be seen from the chart, considerable economic benefits can be achieved. Generally, the investment and management costs are reimbursed within 5 years ($144 \times 5 = 720$).

Table 2. Economic analysis of PV-Huanghuai goats.

	Facilities & input	Total Investment (10^4 yuan)	Annual income (10^4 yuan)
Agricultural facilities	Sheepcote	1.5	From Huanghuai goats: 15
	Land rent	1.8	
	Sheep and feed	20	
	Management and maintenance	0.96	
PV Facilities	Photovoltaic facilities	10	From generated energy: 0.7
	Total	34.26	15.7

Considerable economic benefits can be achieved, with reimbursement of investment and management costs within 2.5 years ($15.7 \times 2.5 = 39.25$).

Table 3. Economic analysis of PV-Penaeus vannamei.

	Facilities & input	Total Investment (10^4 yuan)	Annual income (10^4 yuan)
Agricultural facilities	Shrimp pond facilities	1.3	From Shrimp: 25
	Shrimp pond rental	2.3	
	Shrimp seedling and food	20	
	Management and maintenance	1.25	
PV Facilities	Photovoltaic facilities	400	From generated energy: 60
	Total	424.85	857

This solution yields significant economic benefits, enabling to reimburse the investment and maintenance costs within 5 years ($85.7 \times 5 = 428.5$).

Table 4. Economic analysis of PV-tea.

	Facilities & input	Total Investment (10^4 yuan)	Annual income (10^4 yuan)
Agricultural facilities	land rental	5	from tea: 22.5
	Tea seedling cost	20	
	Management and maintenance	2.5	
PV Facilities	Photovoltaic facilities	125	from generated energy: 19
	Total	152.5	41.5

The PV tea garden offers considerable economic benefits. The costs are reimbursed within 4 years ($41.5 \times 4 = 166$).

4.1.2 *PV livestock breeding*

For example, in the case of a pasture covering an area of 1ha which is used for breeding 100 Huanghuai goats and sheep in a shed employing roof layout of PV panels, the installed capacity of PV power generation is 10 KW.

4.1.3 *PV fisheries*

A pond for farming shrimp (Penaeus vannamei), with the area of 1 ha covered in 60% by PV panels, is characterized by solar power generation capacity of 300 KW.

4.1.4 *PV tea*

As an example, Biluochun tea is planted on the 1ha plot, 20% of tea garden area is covered with, generating 100 KW.

4.2 *Social benefits*

Promoting the PV agriculture project is an important basis for implementing the five development concepts of innovation, coordination, green, openness and sharing, and the implementation of ecological civilization. PV agriculture greatly improved the utilization of land and solar radiation. In addition, the development of PV agriculture can promote local employment, whereas the PV power plant project development, creation of new workplaces, and the development of the tertiary industry will play a significant role in the effective promotion of agricultural restructuring and improvement.

In addition, PV agriculture can provide high-quality agricultural products for the community, also improving the quality of life. The residents can make tours to the PV agriculture facilities as part of sightseeing, also for the promotion and utilization of high-tech in modern agriculture.

4.3 *Ecological benefits*

Solar energy, being a clean energy used in PV power generation, greatly reduces the use of fossil fuels such as coal, enabling to effectively control the emission of carbon dioxide and other greenhouse gases and achieve green, low-carbon, and recycling development.

First of all, PV power generation is conducive to saving non-renewable resources, a single supply of energy balance. The development and construction of PV power plants can effectively reduce the consumption of conventional energy resources, especially coal and petrol resources, protect the ecology, environment and balance the supply of energy.

Secondly, PV power generation can reduce greenhouse gas emissions and the greenhouse effect. It can also help in protecting the environment and reducing the severity of global warming. Taking the Sheyang Wanrun 20 MW PV-fishing project as an example, the average annual grid-connected power generation can reach the capacity of 21.06 million kWh. The project can save 6,381 tons of standard coal per year. For thermal power generation of 1 kWh, the average annual consumption of standard coal is 0.00035 tons, and burning one ton of standard coal emits 2.6 tons CO_2. Therefore, the project is expected to reduce the emission by 24,000 tons of CO_2, 198 tons of SO_2 and 150 tons of NOx per year (Jiang, 2016). Thus, the intensity and significance of energy-saving, emission reduction by PV power plants is very important for enterprises, countries and society as a whole.

5 PROSPECT FOR FUTURE DEVELOPMENT AND RESEARCH

5.1 *Development trends*

High-tech intelligence

The establishment of high-tech intelligent greenhouses will be controlled by the computer terminal, ensuring stability of PV power generation and efficiency of agriculture. The construction of smart micro-network, the formation of energy support for agricultural high-end system environment, may contribute to solving the issue of the lack of electric power in remote areas.

Maximizing benefits

The ultimate goal of PV agriculture still needs to focus on agriculture, farmers, rural areas, which can fundamentally achieve maximum efficiency. PV Agriculture can be combined with agricultural tourism, ecological-tourism and other projects, including the planning of scientific crop cultivation and fish farming, and ultimately promote agricultural development, increase farmers' income and improve the rural landscape.

Mode optimization
China's PV agriculture undergoes transition from the early stages of development to maturity. At this stage, under the support of technology, personnel and theoretical research on PV agriculture will be more systematic, the study of various models will be more in-depth. The models should be continuously studied in order to achieve optimal solutions.

5.2 *Research priorities*

Research on optimization of comprehensive benefits
China's PV agriculture is still in the initial stage, ideal model is still being investigated in order to meet various conditions of particular regions. Both PV and agriculture have to achieve their best performance, in order to improve the economic and employment efficiency, as well as meet the requirements of the ecological and environmental construction.

The issues of how to optimize the layout of PV panel based on agricultural growth patterns, how to realize the PV of ordinary greenhouses, how to determine the matching of cultivation and PV capacity, effectively balance the benefits of PV power generation and agriculture, and maximize the economic, social and ecological benefits of different PV agricultural models need to be considered.

Breakthrough in key technologies, integrated innovation and system supporting
Different PV agricultural models should coordinate with agriculture, to ensure that PV panels are not for sunshine with crops, livestock and poultry. At present, there are certain issues including component spacing and unreasonable installation degree that affect the agricultural biological growth and should be taken into consideration.

The perfect integration of agriculture and PV is realized through setting up PV panels in a reasonable way and establishing corresponding technology sharing platform model to combine modern high-tech planting or breeding with PV power generation.

Standardization of production and management
There is no unified standard for PV agriculture as a modern agricultural science and technology complex for PV power generation and agriculture. In order to avoid the competition of PV agriculture and promote the healthy and orderly development of PV agriculture, it is high time to create a standardized management.

However, the establishment of PV agricultural standards requires a certain basis. On the basis of the demonstrated examples it can be concluded that scientific and technological innovation platform construction, personnel training, improvement of the management system, establishing organizations and test system, integrating industry development, supporting policy etc. still need to be studied further.

5.3 *Difficulties and suggestions*

The introduction of training professionals, the establishment of specialized research institutions, for example, to set up the "Research center for PV agriculture of China" in Nanjing could be considered.

The key hurdle to the integrating the innovation of PV agricultural technology lies in the lack of professional research talents in this field. This issue may be solved through the introduction of foreign and domestic training courses, inviting specialists from Japan, Europe and other countries, or sent trainees to countries with advanced PV agriculture in order to introduce higher technologies to guide China's PV agriculture. Relevant research project related to setting up specialized PV agricultural departments and disciplines in Chinese Universities should be carried out. Training innovative and practical specialists through cooperation with research institutes and related enterprises would improve the technical level.

Improve the PV agricultural standard system

PV agricultural standards should be modular, and based on the status of China's PV agriculture development of various regions. They should consider different models based on successful cases explore the relevant technical data, and summarize the operation and classification of PV agriculture. Operators of mature PV agricultural facilities should sum up their experience, brainstorming the ideas. Professionals of the PV industry should support and analyze the technical points of PV and agricultural integration, in order to improve the standard system and provide a theoretical basis.

Research and strengthen the theory of PV agriculture

PV agriculture is almost in a blank slate in theoretical research. By setting up a disciplinary system, PV agriculture is put forward to a high level of theoretical research. In a successful case, the basic theory of PV agriculture will be summarized. The development of different models of PV agriculture in various regions, the construction of PV agriculture database can be carried out through the data analysis of the research related to PV agriculture.

REFERENCES

Bielińska, E.J., Futa, B., Baran, S., Żukowska, G., Pawłowska, M., Cel, W. & Zhang, T. 2015. Integrating Role of Sustainable Development Paradigm in Shaping the Human-landscape Relation. *Problemy Ekorozwoju/Problems of Sustainable Development*, 10(2):159–168.

Chakraborty, S., Sadku, P.K. & Goswami, U. 2016. Barriers in the Advancement of Solar Energy in Developing Countries like India. *Problemy Ekorozwoju/Problems of Sustainable Development*, 11(2):75–80.

Cheng, S. 2015. PV agricultural standards to be modular, fine [N], China Energy News, 21.

China Photovoltaic Agriculture Working Committee, 2016. The 5th China Photovoltaic Agriculture Standard Seminar and the publication of six PV agriculture standards [EB/OL]. (2016-07-23) [2016-11-18].

Fang, Y., Huang, S., Qin, S. et al. 2015. Analysis of Current Situation and Prospect of PV Agriculture. *Changjiang Vegetables* 18: 35–40.

Gou, H. 2013. Solar photovoltaic power generation and agricultural greenhouses planting combination. *Journal of Henan science and technology*, 2014(10):174–175.

Hanergy Holding Group Limited, 2015. PV agriculture in the ascendant, the correct grasp of the direction is the key. *Agricultural Engineering Technology: Greenhouse gardening* (8):70–72.

http://www.cpvac.org/wangzhangonggao/2016-07-23/2347.html

Huasheng, Q. 2015. Green Energy Technology Co., Ltd .A kind of PV tea tree planting method [P]. Chinese patent: CN104272946 A, 2015-01-14.

Jiang, G., Xu, P., Liu, H. et al. 2016. Present Status of PV Industry and Its Application in Agriculture [J]. *Anhui Agricultural Sciences*, 44 (20): 60–62.

Jiang, F. 2016. Comprehensive evaluation of photovoltaic power generation projects: "Complementation between fishery and light" photovoltaic power plant as an example [D]. Hubei: Hubei University of Technology.

Liu, W. 2014. Development of Photovoltaic Agriculture. *China Rural Science and Technology*, 8: 54–55.

Liu, H. 2015. Biofuel's Sustainable Development under the Trilemma of Energy, Environment and Economy. *Problemy Ekorozwoju/Problems of Sustainable Development*, 10(1):55–59.

National Energy Board of China, 2016. Photovoltaic power generation construction and operation information in the first quarter of 2016[EB/OL]. (2016-04-22)[2016-11-18]. http://www.nea.gov.cn/2016–04/22/c_135303838.htm.

Wang, L. 2012. Several important theoretical problems in the cultivation of edible fungi. *Edible fungus special issue review* 3: 1–3.

Yang, Y., Cao, Y. & Wang, M. 2015. Analysis of the influence of PV agricultural powerhouse project on ecological agriculture. *Energy and Energy Conservation* 2: 73–75.

Yiming, 2013. The country's first photovoltaic roof "pig farm" built. *Northern animal husbandry,* 8:22–22.

Zhang, X., Cui, S. & Liu, F. 2015. Development of PV agriculture. *Anhui Agricultural Sciences* 43(19): 229–231.

Zhao, Y., Meng, X. & Wang, J. 2015. Exploring the development of PV agriculture based on the complementary of fishing and light in Ezhou 20MWp agricultural PV demonstration garden. *Anhui Agricultural Sciences*, 43(22): 360–362.

Żelazna, A. & Gołębiowska, J. 2015. The Measures of Sustainable Development- a Study Based on the European Monitoring of Energy-related Indicators. *Problemy Ekorozwoju/ Problems of Sustainable Development*, 10(2):169–177.

Biofuels and the environment

M. Pawłowska & L. Pawłowski
Lublin University of Technology, Lublin, Poland

ABSTRACT: One of the major threats faced by the modern world is the growing consumption of energy resources. The concerns related to climatic changes, caused by the rising CO_2 emission from the combustion of fossil fuels, increase the interest in a wider application of renewable energy sources. The popular belief is that liquid biofuels constitute one of the important sources of renewable energy. The paper attempts to prove that this belief is erroneous. Wide application of biofuels threatens the sustainable development of the modern world. The authors show that production of biofuels from agricultural crops often violates the sustainable development paradigms, because it limits the access to food by allocating the agricultural land for the cultivation of biofuel crops. Moreover, it also has a negative impact on the environment by reducing the biodiversity and polluting water, while usually having no effect on the mitigation of CO_2 emission.

Keywords: biofuels; renewable energy; carbon dioxide emission; environment

1 INTRODUCTION

A rapid economic development occurred in the 20th century in most parts of the world, which in conjunction with growing human population caused a dramatic increase in the consumption of all planet resources and degradation of the environment. The first report of the Club of Rome of 1972 constituted the first major warning, which drew the attention to the exhaustion of finite Earth resources, indicating the need for correcting the sustainable development paradigms of human civilization (WCED report 1987).

Growing awareness of depleting Earth resources which are essential for the functioning of the human civilization led to the creation of the sustainable development concept formulated for the first time in the report published by the World Commission on Environment and Development in 1987. The most important paradigm mandates that people should act in the way that enables to satisfy their needs without compromising the ability of future generations to satisfy theirs. This provision especially pertains to the quality of environment, preventing from its degradation and the complete depletion of natural resources.

In the following years, the concept of sustainable development has spread into virtually all areas of human activity, setting the directions for the development of civilization (Pawłowski 2013, Papuziński 2013, Cao et al. 2011, Mroczek et al. 2013). Referring to the concept of sustainable development, the UN formulated two important programmes. The United Nations Millennium Development Goals were established in 2000. They drew the attention to the need for such changes in the world, which would secure both the intergenerational equity, related to the necessity of preserving the resources and the quality of the environment so that the future generations could live, as well as the intragenerational equity, allowing the contemporary generation to satisfy their basic needs, including equitable access to the rudimentary resources of the Earth, especially food and water.

In order to better illustrate the issue, the natural resources should be divided into renewable and non-renewable ones. The former comprise the resources which are constantly being replenished in the course of natural processes. This mainly includes the biomass and other resources of animate nature. These resources are renewable, providing that their exploitation does not exceed the rate of production, e.g. in the case of forests it means that the number of cut trees should not be greater than the rate of natural growth. The same is true in the case of animals. For instance, overfishing in seas and oceans has become a problem in recent years, leading to a decline of fish population and irreversible losses.

Typical non-renewable sources include fossil fuels. They were created in the course of millions of years as a result of organic matter transformations, which contained solar radiation energy accumulated through photosynthesis. Industrial exploitation of fossil fuels started in the 18th century and already after two centuries, there is a looming threat of their exhaustion within the lifespan of the current human generation.

It is estimated that the reserves of crude oil will be sufficient for 40–50 years, natural gas—for about 60–70 years, and coal—for another 140–150 years. Although the problem of fossil fuel depletion is important, more attention is devoted to the global warming, which is caused by an increase in the CO_2 concentration in the atmosphere resulting from the combustion of these fuels. The reports of Intergovernmental Panel on Climate Change (IPCC – 2014, World Energy Council 2014) forecast that if the combustion of fossil fuels does not stop, the consequences for the climate will be dire. One of the most important ones includes disruption of rainfall patterns in particular climatic zones, which will negatively impact the food production. Although it is commonly assumed that the effects of climatic changes will be disastrous, one should consider the works of American climatologist Richard Lindzen (Lindzen 2010), bearing in mind that the cost related to the implementation of low-carbon technology will be enormous. Lindzen questions the scale of climate changes predicted by IPCC. This belief is important, because the technologies of CO_2 emission mitigation are connected with additional power consumption, leading to even faster exhaustion of already limited fossil fuel reserves. This, in turn, violates the intergenerational equity paradigm.

Focusing the energy policy solely on mitigating the CO_2 emission threatens the sustainable development of the world. One of the essential paradigms of sustainable development is the intragenerational equity which mandates ensuring equitable access to basic, necessary goods, including food, for all people. Despite the appeals, almost 1 billion people still starve. It turns out that adoption of the EU 2009 directive requiring 10% share of biofuels in transportation till 2020 may threaten the intragenerational equity paradigm of sustainable development. Simultaneously, it has a questionable influence on the mitigation of CO_2 emission, because the production of biofuels causes both an increase of food prices and a decrease of its production (Duran et al. 2013).

Energy is the most important resource for sustainable development. It permeates all aspects of human life: social, economic, environmental ones, including access to water, as well as food production, health, education, and even gender-related issues. Hence, supplying our civilization with energy is one of the key tasks of implementing the sustainable development strategy.

Taking into account the environment, economy, and the social aspects it can be stated that the global energy system is unsustainable. The current trend of energy supply leads to the exhaustion of resources and will probably cause climatic changes. Ensuring energy supply without any negative environmental impacts is an enormous challenge for our civilization.

It is believed that renewable energy sources are an essential element of energy supply. These include, for instance, the solar energy, used both as thermal and electric energy generated in photovoltaic panels, wind energy, power derived from the hydrokinetic energy, geothermal energy, as well as biomass energy. Utilizing biomass as a source of energy is the most controversial solution. Wide application of biomass from agricultural crops for energy production often violates the concept of sustainable development, which we will attempt to demonstrate in this paper.

2 COMPETITION WITH THE FOOD PRODUCTION

The demand for food will continue to grow for two reasons, i.e. growing human population and larger number of better-fed people. Meanwhile, 250 thousand people starve to death each day, and approximately 780 million people in developing countries and 27 million people in developed countries are malnourished (Unicef 2016, Oxfam 2012). In this case, allocating vast areas of land for the cultivation of biomass for fuel purposes is morally dubious. This is especially true in relation to the liquid biofuels used in transportation. In line with the decision of EU Commission made in 2009, as much as 10% of energy used in transportation should come from the biofuels derived from agricultural crops. In order to make biofuels cost-effective, the European governments subsidize the powerful industrial and agricultural lobbies. For instance, in 2020, each person in the United Kingdom will pay about £35 in biofuel subsidies (£1–2 billion in total); in Germany, this figure will roughly reach €30 (€1.4–2.2 billion in total).

In the United States as well, the production of ethanol developed steadily owing to subsidies. It was mainly produced from corn and used as a fuel additive (EIA 2012). In 2011, as much as 127 billion tons of corn was allocated for the production of bioethanol, which constitutes 40% of annual production. The production of ethanol for fuel purposes required $6 billion dollars of governmental subsidies. Allocating such high amounts of corn for the production of ethanol resulted in a doubling of corn prices. Huge volume of agricultural crops import for biofuel production by the European Union caused a dramatic, 2.5-fold increase in the FAO food price index.

An increase in food prices is especially felt by the poor people, who spend an increasing share of their income on sustenance.

In view of the above-mentioned data, utilizing biomass from agricultural crops as a source of energy threatens the realization of sustainable development because it violates the intragenerational equity paradigm by limiting the access to food for the poor.

3 INFLUENCE OF BIOFUELS ON THE EMISSION OF GREENHOUSE GASES

Promotion of biofuels is based on an erroneous assumption that the plants only release the same amount of CO_2 during combustion, as was absorbed throughout their growth. This simplified way of thinking does not account for the entire production cycle of biofuels. These estimations do not consider the changes in land use, as well as the energy outlays required for the cultivation and processing of biomass into biofuels.

Strong pressures on the usage of biofuels in transportation resulting from the European Union policy caused that the tropical forests, especially in the developing countries, are cut down and biofuel plants are being cultivated in their place. The studies conducted by Danielsen et al. indicate that the capacity of the tropical forests to absorb CO_2 is much greater than the one of the plants used for biofuel production. Consequently, cultivation of biofuel plants lowers the CO_2 absorption in the areas of cut tropical forests. Transforming the tropical forests and peatlands into biofuel crops leads to an additional emission of CO_2, reaching the amount of approximately 55 Mg CO_2 from 1 ha over the period of 120 years. Hence, the application of biofuels derived from agricultural crops does not mitigate the CO_2 emissions (Melillo et al. 2009).

Moreover, in order to produce a biofuel, e.g. corn bioethanol, it is necessary to spend energy on the cultivation, production of fertilizers, plant harvesting, as well as fuel processing with fermentation and distillation. Using the Life Cycle Analysis method it was shown that the amount of CO_2 emitted per unit of energy derived from corn bioethanol is 60% greater in relation to the amount of CO_2 emitted from the combustion of equivalent amount of crude oil fuels. Even Brazil, where the production of bioethanol from sugar cane is most advanced and fully utilizes the residual biomass, e.g. stalks for the production of thermal energy, failed to lower the emission of CO_2 per unit of energy below the level characterizing the liquid biofuels derived from crude oil (Bullis 2011).

However, in the case of Brazil, the development of sugar cane ethanol production led to the creation of 700 thousand new jobs, which can be considered a positive effect, increasing the social sustainability. This enabled Brazil to become independent from the import of liquid fuels, and the price of energy derived from ethanol is competitive to the one of petrol. Thus, it can be said that Brazil managed to ensure sustainable access to liquid fuels in transportation, albeit without mitigating the CO_2 emission. However, this is an exceptional case (Walter et al. 2011).

Full Life Cycle Analysis showed that cultivation of certain plants for biofuel production, such as rapeseed, requires heavy application of fertilizers, which increases the emission of nitrous oxide, i.e. a greenhouse gas, thus significantly aggravating the problem of global warming. In the case of rapeseed cultivation, the generated nitrous oxide may increase the global warming by as much as 70%.

4 EVALUATION OF BIOFUELS IN THE ASPECT OF ENERGY EFFICIENCY

According to the research carried out by professor Piementel (Pimentel et al. 2009), the amount of energy used for the production of bioethanol is higher than the energy gained from the combustion of ethanol in automobile engines.

The process of corn ethanol production consumes 29% more energy than is gained from the combustion of ethanol derived from grass – by 45% – and from wood – by 57%.

Similar situation occurs with soy biodiesel production, which consumes 27% more energy than is gained from the produced biodiesel, whereas in the case of the biodiesel produced from sunflower seeds, this value is increased to 118%.

The above-mentioned data shows that in the United States (and probably other developed countries as well) the cultivation of plants for liquid fuel production is not sustainable, because it increases the consumption of fossil fuels and CO_2 emission even further.

5 IMPACT OF BIOFUELS ON THE ENVIRONMENT

Expansion of areas for biofuel cultivation destroys the environment and biodiversity.

More than the half of terrestrial animals lives in the tropical forests. The forests in south-eastern Asia, which are highly abundant in habitats of various organisms, are most threatened by the creation of biofuel plantations. Tropical forests also absorb approximately 46% of carbon dioxide contained in the atmosphere. Their destruction may lead to a 25% increase of carbon dioxide content in the atmosphere.

Therefore, there is an internal contradiction related to the allocation of the areas occupied by tropical forests for the cultivation of biofuel plants with low emission of carbon dioxide. It is estimated that converting forests into biofuel plantations would reduce the number of species inhabiting these areas by the factor of five.

Production of liquid biofuels for transportation also exerts a negative influence on the aquatic environment due to a heavy consumption of water used both for the irrigation of crops and in the processing of plants into biofuels. Notably, processing of plants yields substantial amounts of wastewater characterized by high environmental impact, for instance, producing 1 litre of ethanol simultaneously creates 6–12 litres of highly polluted wastewater. Meanwhile, the shortage of water already negatively impacts the food production.

Generally, producing 1 litre of bioethanol consumes approximately 2500 litres of water, which is equivalent amount to the one required to produce food for one person. In order to irrigate 30 000 000 hectares used for the cultivation of biofuel crops, 180 km^3 of fresh water will be needed (Nylor et al. 2007, Escobar et al. 2009).

It should be noted that due to the growth of population to 8.3 billion in 2030 (7.2 billion in 2012), the demand for food, water, and energy will continue to increase by 35%, 40%, and 50%, respectively.

Vast areas of monocultures, usually employed in the case of biofuel crops, require the application of substantial amounts of herbicides and pesticides, which subsequently infiltrate into the groundwater, polluting it. The soy crops in Brazil are an example of a negative impact of pesticide usage. Wide application of pesticides and herbicides threatens the wetlands of Pantanal, an important region which provides habitats for hundreds of bird, mammal, and reptile species. Another example includes a 20 000 ha sugar cane plantation allocated for ethanol production, located in the Tana River Delta in Kenya. The planned water consumption of 1680 m^3 of water/min constitutes about 30% of river flow, seriously threatening the local ecosystem, inhabited by 345 species of water and marsh birds.

6 CONCLUSIONS

At present, the supply of primary energy is being urgently sought to mitigate the CO_2 emission and climatic changes. However, majority of biofuels do not meet the criteria of sustainable development. Their application, in most cases, does not lead to the mitigation of CO_2 emissions. Moreover, in some cases, the energy gained from the combustion of biofuels is lower than the energy consumed for their production. In tropical regions, the plantations of biofuels often exert a negative influence on the local environment. The use of biofuels, mainly by the European Union member countries, contributes to the rising food prices. Thus, it threatens the intragenerational equity, one of the basic paradigms of sustainable development. Only the biofuels produced from waste biomass or the biomass cultivated on depleted soil may be sustainable and mitigate the CO_2 emission to a certain extent.

It should be taken into consideration that every human activity has a negative impact on the environment – even the process of breathing emits CO_2. The CO_2 emission is an important, but not paramount factor influencing the development of human civilization.

However, the increase of CO_2 concentration in the atmosphere can also be beneficial. It contributes to a faster growth of biomass, including food, as its rate of growth is dependent on the assimilation of CO_2 from the atmosphere.

Undoubtedly, the supply of energy will be one of the essential factors governing the development or even survival of the human civilization.

A critical analysis of the current trends in energy supply seems to indicate that it is necessary to utilize all available energy sources, simultaneously conducting works aiming at mitigating their impact on the natural environment.

The above-mentioned examples show that ensuring adequate energy supply becomes one of the most important problems faced by the world. Solving it will require a comprehensive approach to the assessment of all primal energy sources.

REFERENCES

Bullis, K. 2011. *Do Biofuels Reduce Greenhouse Gases*? MIT Technology Review.

Duran, J., Golusinm M., Ivaovic, O.M., Jovanovic, L. & Andrejevic, A. 2013. Renewable Energy and Socio-economic Development in the European Union. *Problemy Ekorozwoju/Problems of Sustainable Development*, 8(1):105–114.

EIA, 2012, How much ethanol is produced, imported, and consumed in the United States?

Escobar, J.C., Lora, E.S., Venturini, O.J., Yáñez, E.E., Castillo, E.F. & Almazan, O. 2009. Biofuels: Environment, technology and food security. *Renewable and Sustainable Energy Reviews* 13: 1275–1287.

IPCC-2014, Intergovernmental Panel for Climate Change Report: Mitigation of Climate Change.

Lindzen, R.S. 2010. Global Warming: The Origin and Nature of the Alleged Scientific Consensus. *Problemy Ekorozwoju/Problems of Sustainable Development*, 5(2): 13–28.

Melillo, J.M. et al. 2009. Indirect Emissions from Biofuels: How Important? *Science* 326, 1397; DOI: 10.1126/science.1180251.

Mroczek, B., Kurpas, D. & Klera, M. 2013. Sustainable Development and Wind Farms. *Problemy Ekorozwoju/Problems of Sustainable Development*, 8(2): 113–122.

Naylor, R.L., Liska, A., Burke, M.B., Falcon, W.P., Gaskell, J.C., Rozelle, S.D. & Cassman K.G. 2007. The Ripple Effect: Biofuels, Food Security, and the Environment". Agronomy & Horticulture—Faculty Publications, Paper 386.

Oxfam Briefing Paper, 2012, The Hunger Grains Briefing Paper, www.oxfam.ca/news-and-publications.

Papuziński, A. 2013. The Axiology of Sustainable Development: An Attempt at Typologization, *Problemy Ekorozwoju/Problems of Sustainable Development*, 8(1): 5–25.

Pawłowski, A. & Cao, Y, 2013, The role of CO_2 in the Earth's ecosystem and the possibility of controlling flows between subsystems. *Gospodarka surowcami mineralnymi – Mineral resources management*, 30(4): 5–20.

Pawłowski, A. 2013. Sustainable Development and Globalization. *Problemy Ekorozwoju/Problems of Sustainable Development*, 8(2): 5–16.

Piementel, D., Marklein, A., Toth, M.A., Karpoff, M.N., Paul, G.S., Mccormack, R. et al. 2009. Food versus biofuels: Environmental and economic cost. *Hum Ecol*. 37:1–12. DOI: 10.1007/s10745-009-9215-8

Udo, V. & Pawłowski, A. 2011. Human Progress Towards Equitable Sustainable Development – part II: Empirical Exploration, *Problemy Ekorozwoju/Problems of Sustainable Development*, 6(2): 33–62.

Unicef, 2016, 2016 World Hunger and Poverty Facts and Statistics.

Walter, A., Dolzan, P., Quilodrán, O., de Oliveira, J.G., da Silva, C., Piacente, F. & Segerstedt, A. 2011. Sustainability Assessment of Bio-ethanol Production in Brazil Considering Land Use Change, GHG Emissions and Socio-economic Aspects. *Energy Policy* 39: 5703–5716.

World Energy Council, 2013, World Energy Resources 2013 Survey.

Pretreatments to enhance the digestibility of recalcitrant waste—current trends

A. Montusiewicz & M. Pawłowska
Lublin University of Technology, Lublin, Poland

ABSTRACT: Efficient degradation of recalcitrant waste requires using pretreatment prior to anaerobic digestion of the substrate. Pretreatment is a prerequisite method for reducing structural and compositional impediments of residues, particularly in terms of lignocellulosic and keratin-rich biomass. As a result, the compounds become more accessible to enzymes, which significantly improves their conversion. This paper shortly reviews the current trends related to the techniques used for pretreatment of recalcitrant waste that enhance its digestibility. In the consecutive sections, different pretreatment methods (physical, physicochemical, chemical, biological as well as combined one) are described and their effects on improving the enzymatic hydrolysis of complex structures are discussed.

Keywords: pretreatment; recalcitrant waste; lignocellulosic biomass; keratin-rich waste; enhanced digestibility

1 INTRODUCTION

The concern about using organic waste as renewable energy sources is currently a critical worldwide issue. The recommended "waste-to-energy" approach fulfils both the demands of sustainable energy policy (reducing the dependency on fossil fuels and increasing energy efficiency) and the environmental requirements (including mitigation of CO_2 emission and related global warming). Accordingly, producing energy from waste becomes a promising alternative that may be classified as a cost-effective resource recovery process (Kondusamy and Kalamdhad, 2014). In this area, anaerobic digestion (AD) is commonly known to be a suitable treatment that produces an energy carrier (i.e. biogas composed mainly of CH_4 and CO_2) and reduces waste. Another benefit includes generation of digest with high nutritional potential for agriculture.

Anaerobic digestion (AD) consists of complex biochemical reactions carried out by numerous microorganisms which degrade organic compounds in oxygen-free conditions. The process involves four stages: hydrolysis, acidogenesis, acetogenesis and methanogenesis, showing high sensitivity to several factors. They include the type and chemical composition of the substrate, environmental and operational conditions (i.e. temperature, pH, organic loading rate, hydraulic retention time), and the concentration of inhibitory compounds (Montusiewicz, 2008). It is widely known that organic waste varies in terms of its structure and composition, particularly with respect to its biodegradability and bioaccessibility. Depending on the sources (municipal, agricultural, food and industrial wastes), specific organic substances may predominate. Hence, the classification involves: carbohydrate-rich materials (e.g. food waste, lignocellulosic residues from agriculture and forestry), protein-rich materials (e.g. slaughterhouse waste, pig and chicken manure, dairy whey) and fat-rich materials (e.g. waste from olive oil industry) (Hagos et al., 2017). Each of these requires an adequate approach to treatment and may cause some technological problems resulting in a disruption or even failure of the anaerobic system. For example, high content of easily degradable monosaccharides (typically present in food waste) may result in rapid accumulation of volatile fatty acids, the subsequent drop of the pH value and the suppression of methanogenesis. Decomposition of protein-rich waste leads to the release of high concentrations of ammonium ions/ammonia, which can strongly inhibit methanogens. High content of lipids may result in blocking, adsorption into biomass, foaming and inhibition of microbial activity by long-chain fatty acids (Wagner et al., 2013).

Different strategies are recommended in order to overcome the disadvantages attributable to mono-substrate anaerobic digestion (Figure 1). They include: waste pretreatment (P), two-stage anaerobic digestion (TSAD), anaerobic co-digestion (AcoD) and bioaugmentation (B). Some of them alter the structural characteristics of waste enhancing its accessibility and digestibility, the others improve nutritional balance, metabolic properties and operational factors facilitating synergistic effect of microorganisms. Regardless of the strategy chosen, an enhanced biogas production may be obtained. However, both the specific substrate characteristics (particularly its complexity and chemical structure) and cost-efficient analysis should be taken into account while selecting an optimal solution.

Waste recalcitrance, which significantly limits the AD efficiency, is attributable to the heterogeneous structure and physicochemical complexity of the compounds included. This concerns primarily lignocellulosic biomass which is the most abundant organic material that can be used as a source of renewable energy. Additionally, keratin-rich waste should be involved. Lignocellulosic matter is hardly digestible because of its complex structure that comprises matrix polymers (hemicellulose, pectins, and lignin) surrounding the cellulosic microfibrils in the plant cell wall (Himmel and Picataggio, 2009). Keratin-rich waste (i.e. feathers, hair, wool, horns) contains insoluble structural proteins with the polypeptides chains that are tightly packed and highly cross-linked with disulfide bridges, hydrogen bonds, and hydrophobic interactions. This structure makes keratin mechanically stable and resistant to enzymatic attack (Daroit et al., 2009).

Efficient degradation of recalcitrant matter requires pretreatment prior to anaerobic digestion as a prerequisite method of reducing structural and compositional impediments of the substrate (Sun et al., 2016). This enables to increase decomposition rates (mainly hydrolysis) and enhances both biogas yields and productivity, since pretreatment makes the compounds more accessible to enzymes and thus improves their conversion (Wyman et al., 2005). With regard to cellulosic waste, the pretreatment is aimed at breaking the lignin layer that protects the cellulose and hemicellulose, decreasing the cellulose crystallinity and increasing the biomass porosity. Keratin-rich waste, by contrast, requires the destruction of the cross-linking

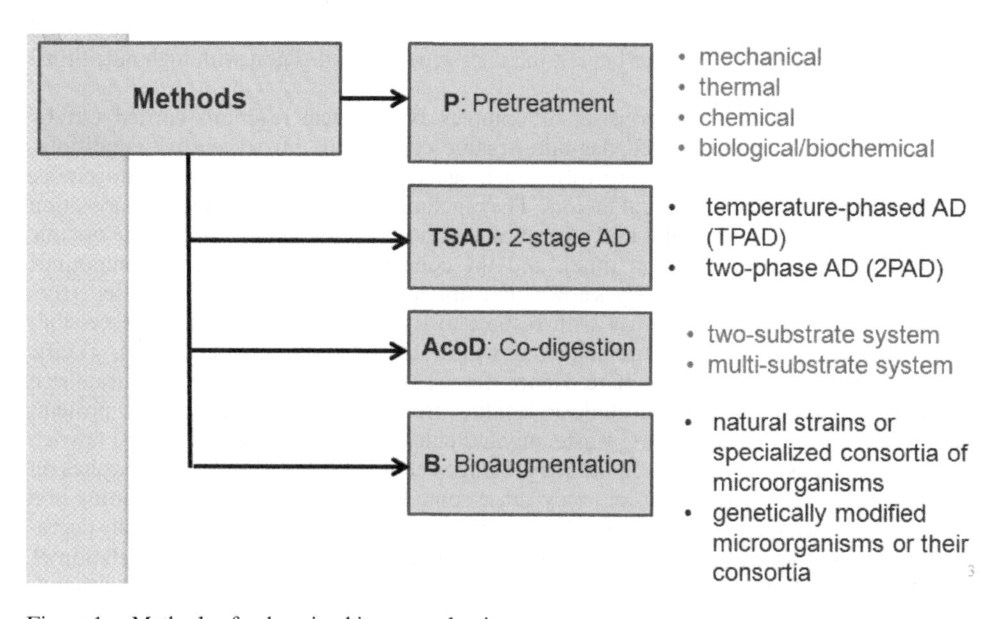

Figure 1. Methods of enhancing biogas production.

between the polypeptides chain prior to its anaerobic digestion (Patinvoh et al., 2017). While choosing an adequate pretreatment, physicochemical characteristics of the substrate as well as the advantages and challenges of different methods have to be considered. Importantly, an excessive hydrolysis must be avoided to prevent the AD system from both overloading by intermediate products and deceleration of methanogens activity. For example, the effective pretreatment of lignocellulosic material should improve the formation of sugars, avoiding the degradation of monosaccharides derived from carbohydrates and minimizing the formation of inhibitors (i.e. phenolic compounds) for the subsequent AD stages. Moreover, it should be environmentally friendly and characterized by low capital and operational costs, as well as low energy demand (Bhutto et al., 2017).

This paper shortly reviews the recent progress in pretreatments used for recalcitrant waste prior to its anaerobic digestion. In the following sections, pretreatments are divided into four categories to discuss their effects on improving the enzymatic hydrolysis of complex structures such as lignocellulosic compounds and keratin.

2 PRETREATMENT PROCESSES

To date, various pretreatments have been investigated and recommended for different types of recalcitrant waste. Current classification comprises physical, physicochemical, chemical, and biological methods (Mupondwa et al., 2017; Bhutto et al., 2017). All of them lead to overcoming the limiting step of substrate hydrolysis, although selection of the suitable strategy and integration of pretreatment methods differ depending on downstream process. It is worth noting that some methods dedicated to bioethanol production have not been evaluated in AD for biogas generation. Moreover, pretreatment conditions for AD can be milder than those for bioethanol production (Zheng et al., 2014). Thus, the choice of pretreatment is essential regarding its impact both on the efficiency of the subsequent processes and the overall treatment costs.

3 PHYSICAL PRETREATMENTS

Physical pretreatments are aimed at increasing the accessible specific substrate area known as a crucial factor that affects enzymatic hydrolysis of organic matter (Alvira et al., 2010). Moreover, these improve particle size distribution and densification, increase bulk density and porosity, and enhance flow property. The physical methods include mechanical comminution (chipping, grinding and milling), irradiation (gamma rays, electron beam, microwave) and ultrasonic pretreatment (acoustic and hydrodynamic cavitation). However, their mechanisms and effects are somewhat different. Comminution is typically used for reducing the particle size and disrupting structural regularity of waste (i.e. cellulose crystallinity). Irradiation modifies polymer materials through degradation, grafting and cross-linking (Bhutto et al., 2017). However, it rather assists other kinds of treatment, enhancing its efficiency. Ultrasonic pretreatment alters both the structure of waste and its chemical composition because of the mechanoacoustic/hydrodynamic and sonochemical effects that occur inside a liquid when subjected to changes in the pressure field over time and distance. These cavitational effects are generated by ultrasound acoustic waves with the frequency range from 10 kHz to 100 MHz or by the passage of the liquid through constrictions (orifice plates, curved channels, Venturi or throttling valves) inducing local supercritical conditions that include temperatures in the order of 10,000 degrees Kelvin and pressures up to a GPa. Consequently, the mechanisms such as thermal decomposition, shockwaves, shear forces, pressure gradients, as well as the reaction of hydroxyl radicals and reactive hydrogen atoms with the pollutants may play a significant role (Arrojo & Benito, 2008).

While using mechanical comminution, different machines can be involved in production of various particle sizes. Chipping yields the fraction of 10–30 mm, whereas milling/grinding ensures minor size of 0.2–2 mm. Importantly, a proper choice of device is decisive for subsequent AD efficiency, because an excessive reduction of particle size may result in decreasing biogas production and lead to overproduction of volatile fatty acids. The associated devices comprise: ball, vibro, hammer, knife, two roll, colloid, and attrition mills as well as extruders. Their selection depends on the moisture content of waste; for example, colloid mills and extruders are suitable only for materials with moisture content of more than 15–20% (Zheng et al., 2014). Among recent studies in this area, Jiang et al. (2017) investigated mechanical milling of wood and evaluated the effects of input moisture content as well as substrate particle size on both the characteristics of micronized matter and the specific energy consumption. The researchers used a vibratory ring and puck mill for effective comminution of wooden particles to < 20 μm within several minutes. It was found that lower moisture content resulted in much rounder particles with lower crystallinity, whereas the micronized particles were generated at higher moisture revealing larger crystallinity. The changes of particle size were highly correlated to the specific energy consumption through milling. Input moisture content and particle size affected the energy intensity of grinding woody biomass. Interestingly, a multi-step milling was recommended as a cost-effective strategy for achieving fine woody biomass.

In terms of irradiation, microwave heating has been currently investigated by some researchers. Pecorini et al. (2016) studied microwave and autoclave pretreatments of two synthetic organic fractions of municipal solid waste (OFMSW) with different lignocellulosic contents. Both pretreatments were found to hydrolyze a significant fraction of recalcitrant organic matter which was revealed by an increase of soluble chemical oxygen demand (SCOD) up to 219.8%, comparing to untreated substrates. However, most of the SCOD obtained via pretreatments was composed of non-biodegradable substances. Hence, methane production from the substrates increased by only 8.5% after microwave treatment and by 1.0–4.4% after autoclave treatment (the higher value was achieved for the most lignocellulosic OFMSW). No energy profit was registered due to the low increase in biogas production. Microwave heating was also tested as a pretreatment for lignocellulosic agroindustrial waste prior to its anaerobic digestion (Pellera and Gidarakos, 2017). The substrates included winery waste, cotton gin waste, olive pomace and juice industry waste. While pretreating winery waste and juice industry waste, relatively high solubilization levels were achieved for both materials. In contrast, such a finding was not observed in the case to cotton gin waste and olive pomace. It was reported that microwave pretreatment did not improve methane generation, probably because of the presence of recalcitrant or/and inhibitory compounds being a byproduct of the pretreatment.

Acoustic and hydrodynamic cavitation were examined as suitable mechanical pretreatments for recalcitrant biomass. Madison et al. (2017) investigated these methods with regard to microcrystalline cellulose and raw as well as lime-treated sugarcane bagasse. Acoustic cavitation was found to increase microcrystalline cellulose enzymatic digestibility by 37% comparing to untreated samples. In the case of lime-treated sugarcane bagasse, a significant effect was not observed. The researchers suggested that the amorphous components in biomass (hemicellulose and lignin) probably absorb the shock waves, protecting the crystalline cellulose regions. In contrast, hydrodynamic cavitation increased the enzymatic digestibility of both raw and lime-treated sugarcane bagasse with the best results achieved for lime-treated substrate. Ultrasound technique was also used for Napier grass pretreatment (Nagula & Pandit, 2016). The results showed 18% delignification under operational conditions such as: Temperature – 45°C, pretreatment time – 2 h, frequency – 24 kHz, and power – 100 W. Peces et al. (2015) used ultrasonication for pretreatment of brewery spent grain evaluating the effect of its moisture and particle sizes on anaerobic biodegradability of the substrate. The results indicated the total solids of the substrate (TS) as a significant parameter for pretreatment performance. The experimental optimum values was 100 gTS kg^{-1} ensuring an increase of the methane yield by 14% towards the control (without pretreatment). The influence of hydrodynamic cavitation on the biodegradability of brewery spent grain diluted by mechanically pretreated municipal wastewater was examined by Montusiewicz et al. (2017). Hydrodynamic cavitation was applied at an inlet pressure of 7 bar, using an orifice plate with conical concentric hole (3/10 mm as an inlet/outlet diameter) and maintaining 30 recirculation passes through the cavitation zone. It was found that biodegradability index was enhanced from 0.074 to 0.091 revealing the positive impact of the pretreatment on the substrate biodegradability. Moreover, solubilization of carbohydrates was reported with a related monosaccharides release of 87%. Interestingly, the formation of new compounds that would interfere with the subsequent biological treatment was not observed.

4 PHYSICOCHEMICAL PRETREATMENTS

Physicochemical pretreatment is aimed at modifying both the physical and chemical properties of the biomass. Increase in porosity, internal surface area and decrease in crystallinity index of cellulose, as well as decomposition of polymeric structure of polysaccharides, proteins or lipids make the biomass more susceptible to biodegradation. Many methods that are different in terms of delignification and saccharification efficiencies, reducing saccharides yields, utility costs related to energy and chemicals demands, capital cost, and environmental impacts are known. The following methods of biomass pretreatment can be accounted to physicochemical type: steam explosion, CO_2, SO_2 and N_2 explosion; ammonia fiber explosion; liquid hot water, also called hydrothermolysis; wet oxidation; organosolvent and diluted acid pretreatment; alkaline extraction; supercritical fluid and ionic liquids pretreatment, and low temperature pyrolysis (Bhutto et al., 2017; Patinvoh et al., 2017; Raud et al., 2016; Wang et al., 2016a).

From the environmental point of view, the most preferable methods are based on readily-available as well as non-toxic chemical agents and do not generate waste products. Such methods include the liquid hot water and steam explosion which are based on water, and wet oxidation in which air or oxygen are used as oxidizing agents. Among these, liquid hot water in which the biomass is treated under high pressure for tens of minutes with water at temperatures from 180 to 220°C (Bobleter, 1994) is considered to be the most environmentally-friendly method (Bhutto et al., 2017). Stem explosion method involving the heating of the biomass by high-pressure saturated steam at temperature up to 260°C and pressure up to 5 MPa usually for several seconds is included to lower cost pretreatments (Maurya et al., 2015), while the ionic liquids, organosolvents, and ammonia fiber explosion methods that use the specific chemical agents are the most costly (Zhao et al., 2016).

Due to the different mechanisms of the biomass conversion occurring in particular methods, the obtained results differ in terms of the quantity and quality of products, both the desirable, such as monosaccharides or amino acids, as well as undesirable ones, like phenols, furfural, 5-hydroxymetilfurfural (HMF), levulinic acid and formic acid. It was shown that HMF and furfural completely inhibited the methanogenic activity at 2.0 g/L under thermophilic and mesophilic conditions (Ghasimi et al., 2016). According to the review undertaken by Bhutto et al. (2017), wet oxidation is the most effective method of simultaneous removal of cellulose, hemicellulose and lignin from biomass of corn stover. Diluted acid pretreatment is targeted at solubilizing the hemicellulose, while alkaline and ionic liquid pretreatments are aimed at lignin removal. Despite the high efficiency in solubilization of lignocellulose, wet oxidation and acid pretreatments generate the highest amount of toxic products that may inhibit the activity of fermentative microorganisms. Therefore, the special attention should be paid to the effects of these types of biomass pretreatments on the methane yield.

The application of physicochemical pretreatment to enhance the biogas production from lignocellulose biomass has been a subject of numerous studies that were carried out over several decades. Many types of the biomass, different operating conditions, and various combinations of the methods have been examined up to now. The first studies regarding an improvement of biogas production were focused on alkaline chemical and thermochemical pretreatments of such a biomass as wheat straw, cattle manure, water hyacinth, lantana residue, apple and peach leaf (Hassan Dar & Tandon, 1987; Hashimoto, 1986; Baugh & McCarty, 1988, Patel et al., 1993). The thermally stimulated alkali chemical methods are still being developed. Solé-Bundó et al. (in press) employed thermo-alkaline pretreatment for the biogas production in co-digestion of the mixture of 50% of microalgae and 50% of the straw. The biomass was treated with 10% CaO at 75°C for 24 h. It was stated that pretreatment increased the methane yield by 15% compared to the untreated mixture. A significant increase in biogas production was observed in the case of giant reed biomass treated with NaOH in dose 20 g/L by 24 h. The cumulative methane yield was 63% higher in comparison to the untreated biomass (Jiang et al., 2016). Additionally, the net electrical energy production has increased by 27% in comparison with the biomass which did not undergo pretreatment.

Steam explosion pretreatment of the biomass subjected to anaerobic digestion has also been developed for a long time (Castro et al., 1994). The rapid heating of the biomass for several seconds up to few minutes, by using a high-pressure saturated steam at high temperature leads to high sugar production (Bhutto et al., 2017). The influence of different steam explosion conditions on the methane yield of recalcitrant biomass of reed (Lizasoain et al., 2016), pinewood (Khoshnevisan et al. 2016) and miscanthus (Li et al., 2016) was the subject of the studies that have been carried out recently. Lizasoain et al. (2016) showed that digestibility of reed pretreated at 3.4 MPa, 200°C for 15 min. improved remarkably, and the specific methane yield examined in the batch test increased up to 89%. However, the methane yield visibly decreased under harsher conditions, probably because of the formation of toxic compounds during pretreatment. Up to 50% increase in biochemical methane potential after steam explosion pretreatment (at 1.5 MPa, 198°C for 5–10 min) was observed also in the case of Miscanthus lutarioriparius (Li et al., 2016a). On the other hand, the consequential life cycle assessment (CLCA) performed by Khoshnevisan et al. (2016) for the lignocellulosic biofuel production from pinewood in Sweden revealed that steam explosion applied prior to methane production is less environmentally friendly and consumes more energy than the pretreatment with industrial cellulose solvent, N-methylmorpholine-N-oxide (NMMO).

The AFEX biomass pretreatment with liquid anhydrous ammonia under high pressures applied is another method using explosion disruption of the cell structure. The method is fast and carried out in moderate temperatures. Solid phase biomass can be treated in this process (Bhutto et al., 2017). Thus, no water demand is the important advantage of the process. A significant increase in the concentration of the bioavailable sugars can be obtained after the AFEX pretreatments of the recalcitrant biomass. Abdul et al. (2016) observed an almost 4-fold increase in sugar yield after enzymatic hydrolysis of the oil palm empty fruit bunch (OPEFB) fibers pretreated with AFEX. The release of monomeric sugars during the enzymatic hydrolysis of agave bagasse pretreated with AFEX was about 43% of the initial biomass (Perez-Pimenta et al., 2016). However, to the best of our knowledge, there is no data on the effect of AFEX pretreatment on biomethane production.

Many researchers working on improving the biogas production paid their attention on the methods that use oxidizing-agents. According to Schmidt & Thomsen (1998), wet oxidation enables to obtain up to 100% recovery of cellulose and 60% of hemicelulose from the biomass that become available for microorganisms as such can be converted to the biogas. Lissens et al. (2004) stated that thermal wet oxidation of food waste carried out at temperature of 185–220°C, with the retention time of 15 min., caused up to 2-fold increase of methane yield, but an insignificant increase of methane yield of yard waste was observed under these conditions as well. They also stated that thermal wet oxidation of biowaste which was previously digested in a full-scale biogas plant increased its specific methane yield. It indicates that there is still a significant amount of biodegradable compounds that can be converted to biogas in the digested biowaste. Fox and Noike (2004) proved the efficacy of wet oxidation pretreatment (170, 190, and 210°C, with the retention time of 1 h) as a method of an enhancement of the biogas production from newspaper waste. The highest CH_4

conversion efficiency equal to 59% of the initial total COD was obtained from the newspaper pretreated at 190°C, although the highest removal efficiencies of total COD and cellulose from the waste were achieved at 210°C. A significant increase of the specific methane yield was observed when applying wet oxidation to the perennial crops like miscanthus and willow. It showed that the ratio of energy output to input and of costs to benefits for the whole chain of biomass supply and conversion into biogas becomes higher than for the commonly used corn (Uellendahl et al., 2008).

The use of hydrogen peroxide as an oxidizing agent under alkaline conditions proved to be an efficient pretreatment method for hardly biodegradable olive mill waste. By means of the batch tests Siciliano et al. (2016) showed that a high methane yield of approximately 0.328 L CH4/gCODremoved was obtained on the pretreated waste, while an insignificant biogas production was detectable on raw olive mill residues. Hassan et al. (2016) used hydrogen peroxide (H2O2) for the pretreatments of the corn stover. They showed that H2O2 pretreatment was more effective in enhancement of lignocellulosic digestibility and methane production than alkali pretreatment with Ca(OH)2 and NaOH. The methane yield of the corn stover treated by H2O2 were ca. 66% higher than the untreated corn stover, while in case of biomass treated by NaOH, it was ca. 50% higher. Wang et al. (2016a) have tested lower temperature fast pyrolysis (LTFP) as a new thermo-chemical pretreatment method for the biomass subjected to anaerobic digestion. The corn stover was pretreated in the fluidized bed pyrolysis reactor at 180–220°C. LTFP pretreatment increased the methane production by 18%, as compared to the untreated biomass and the maximum methane potential was obtained at 180°C for the flow rate of a carrier gas equal to 3 m^3/h.

Efficient removal of hemicelluloses from the lignocellulosic matrix that leads to an increase in cellulose accessibility for hydrolytic enzymes has been obtained in hydrothermolysis, also called liquid hot water (LHW) pretreatment. In this process, the water at temperature from 180 to 220°C is applied under high pressure as a reaction medium. The process usually runs for several tens of minutes (Bhutto et al., 2017). According to Zhou et al. (2017) the 200°C hydrothermal pretreatment of Miscanthus allowed a 13.2-fold increase of glucose production, compared with the raw material and reduced the digestion time by 50%. Jiang et al. (2016) have examined the effects of LHW pretreatment of giant reed biomass. Under optimal conditions, i.e. at 190°C for 15 min, more than a 2-fold increase in glucose yield was observed, while the methane production did not increase significantly. These findings suggest that undesired degradation products were generated during pretreatment. Additionally, negative net electrical energy production occurred because the high energy demand.

Another method used for the hemicellulose and lignin solubilization and releasing the cellulose is the pretreatment with organosolvents. In this method, the biomass is treated with a mixture of organic liquid and water (Bhutto et al., 2017). Different chemicals, like methanol, ethanol, acetone or glycols, ethylene glycol, acetic acid, and formic acid have been examined as the lignocellulose solvents (Sindhu et al., 2012; Ostovareh et al., 2015; Zhang et al., 2016). This organosolvent pretreatment is considered as an emerging method of the pretreatment of biomass used for biofuel production that is characterized by numerous advantages. It enables to fractionate the lignocellulose into high purity components, which obviously increases the biomass usability for production of biofuels. Furthermore, the solvents can be recovered and reused, which reduces the operating cost of the process. The significant influence of the organosolvent pretreatment of lignocellulosic biomass on biogas production was showed in the case sweet sorghum stalks. Up to 270% increase of biomethane yield was obtained from the sweet sorghum stalks pretreated with ethanol at 100–160°C for 30 min. It corresponded to 92% of the theoretical yield (Ostovareh et al., 2015). Liquid-to-solid ratio is a major parameter for optimization of the process in terms of reducing its capital costs, due to decrease in water demand that enables to use smaller tanks and pumps for the same quantities of feedstock (Zhang et al., 2016).

The changes in the physical structure and chemical composition can be achieved also by treating the lignocellulosic biomass with ionic liquids. There are salts which comprise organic cations and organic or inorganic anions able to dissolve the cellulose (Zhao et al.,

2016). Disruption of the hydrogen bond in cellulose structure causes a delignification and reduction in cellulose crystallinity which makes the biomass amenable for further microbial degradation (Singh et al., 2009). The effect of ionic liquid pretreatment on the biogas production from the biomass is insufficiently explored. One of the spars study concerns the pretreatment of a biomass of water hyacinth with 1-ethyl-3-methlyimidazolium acetate and 1-N-butyl-3-methlyimidazolium chloride under 120°C for 120 min (Gao et al., 2013). However, the formic acid and levulinic acid, the toxicity of which for anaerobic microorganisms was confirmed by Park et al. (2013), were produced. The highest cumulative biogas production was observed after the pretreatment with 1-N-butyl-3-methyimidazolium chloride carried out at 120°C for 120 min. It was almost 2-fold higher that the value for untreated biomass.

Some recently published papers have been focused on the comparison of the effects of different pretreatment methods on lignocellulose decomposition and methane yield. For example, Bolado-Rodríguez et al. (2016) compared the effect of thermal, acid, alkaline and alkaline-peroxide pretreatments of wheat straw and sugarcane bagasse. They observed the highest methane productions from thermally pretreated biomass. Methane productions for wheat straw and bagasse were 29% and 11% higher compared to the raw materials, respectively. All the pretreatments released formic and acetic acids, as well as phenolic compounds. However, only acid pretreatment led to the production of furfural and 5-hydroxymetilfurfural (HMF).

Xia et al. (2012) showed that keratin-containing materials can be successfully used for biogas production. They observed that total methane production during the co-digestion of ground chicken feathers with swine manure was 130% higher compared to the control sample without the feathers. However, Salminen et al. (2003) showed that the feathers have low methane yield, and the enhancement of its biodegradation in needed. The dissolution of the keratin is the first and essential step of reusing the keratin-rich waste (Wang et al., 2016b). Physicochemical pretreatment can be successfully used for this purpose. Salminen et al. (2003) compared the effects of different pretreatments of poultry slaughtering residues. They stated that combined thermal (120°C, 5 min) and enzymatic (alkaline endopeptidase) pre-treatments allow to increase methane yield even by 51%, while the separately applied thermal (70–120°C, 5–60 min), chemical (NaOH) and enzymatic pre-treatments were less effective, leading to the increase of methane yield by 5 to 32%.

Bhavsar et al. (2106) compared the effect of high temperature alkaline hydrolysis (using KOH and CaO) of wool keratin with the results of superheated water hydrolysis (SWH) treatment carried out at the temperatures of 140° and 170°. They found that both processes led to the production of low molecular weight proteins and amino acids. However, because of the lower cost and environmental impact, superheated steam hydrolysis is the better solution. Different other chemicals were tested for keratin-waste pretreatment. Barone et al. (2006) extracted the proteins from poultry feathers by a mixture of glycerol, water, and sodium sulfite (Na_2SO_3) as processing agents. Sodium sulfite is able to break the S-S cysteine bonds with intra- or intermolecular cysteine molecules (Bhutto et al., 2017), and the keratin contain the large amount of the amino acid cysteine compared with other proteins (Vincent, 1990). Shavandi et al. (2016) have examined the efficiency of keratin extraction from Merino wool by means of five chemical pretreatment types: sulfitolysis, alkali hydrolysis, reduction, oxidation, and extraction using ionic liquid. They stated that the highest protein yield (95% and 89%) was obtained by ionic liquid and sulfitolysis methods while the oxidation method was the least effective. Wang et al. (2016b) have obtained 72% dissolubility of wool keratin by using L-cysteine as a reducing agent. Tonin et al. (2006) have examined the influence of chemical-free steam explosion of keratin fibers. They observed an extent disruption of the fiber structure, reduction of the molecular weight of water-soluble peptides and free amino acids when compared with the original wool. However, the influence of the products obtained in the above mentioned processes have been not examined in terms of the biogas production.

5 BIOLOGICAL PRETREATMENTS

Biological pretreatment is based on fungi, bacterial and enzymatic activity. Several fungi classes such as brown-, white- and soft-rot fungi are used to degrade lignocellulosic matter. The first of these mainly attacks cellulose, others depolymerize both cellulose and lignin (Cheng & Timilsina, 2011). Among fungi, Trichoderma viride has been recently reported as a prolific producer of cellulose degrading enzymes exhibiting tolerance to both high pH and temperature, as well as easy cultivability and high level of genetic diversity. These enyzmes comprise endoglucanases, cellobiohydrolases (exoglucanases) and β-glucosidases and act synergistically (Mutschlechner et al., 2015). Moreover, T. viride produces carbohydrate-active enzymes such as hemicellulases and pectinases. Various white-rot fungi have been used for pretreatment of different lignocellulosic materials (e.g. rice straw, wheat straw, corn stalks, grass hay, wood fiber). They include Phanerochaete chrysosporium, Ceriporia lacerata, Cyathus stercoletus, Ceriporiopsis subvermispora, Pycnoporus cinnabarinus, Pleurotus ostreatus, Pleurotus sajor-caju, Irpex lacteus and others (Du et al., 2011; Castoldi et al., 2014; Rouches et al., 2016). Promising effect has also been reported for white-rot basidiomycete Punctularia sp. TUFC20056 with regard to bamboo culms (Suchara et al., 2012). While degrading the keratin or keratin-like waste (e.g. fibrin, elastin, collagen, gelatin), distinct microorganisms are involved, including Bacillus sp. (Patinvoh et al., 2016) and Aspergillus sp. (Mazotto et al., 2013). These microorganisms effectively hydrolyze both α- and β-keratin, and/or degrade keratinous and non-keratinous proteins. According to Fellahi et al. (2014), the released feather hydrolysate can contain tyrosine, phenylalanine and histidine, while the wool lysate may include aspartic acid, methionine, tyrosine, phenylalanine, histidine, and lysine.

Another biological method involves microbial consortia that exhibit high cellulose- and hemicellulose-degradation ability. In such cases, microorganisms are screened from various ecological niches where specialized consortia are typically formed to degrade rotten lignocellulosic biomass. The ones, which have a high hydrolytic activity and remain biologically stable as well as controllable are the most valuable (Zuroff et al., 2013). In addition, complex microbial agents may be applied, for example the ones constructed in a form of freeze-dried powder containing a mixture of pure strains of yeast and cellulolytic bacteria (Zhong et al., 2011).

Among biological pretreatments, the use of enzymes with great hydrolytic activity is also included. This enables to overcome the constraints of recalcitrant biomass hydrolysis and thus enhances the degradation rates in anaerobic digestion. Moreover, enzymes are safer than chemicals, making this treatment environmentally friendly. Cellulase, hemicellulase and starch-degrading enzymes (i.e. amylases) are most frequently used for pretreatment of various lignocellulosic materials. However, efficient degradation of lignin depends on the lignolytic enzymes produced by basidiomycete like laccase, lignin peroxidases, versatile peroxidases and manganese peroxidases (Sindhu et al., 2016; Mupondwa et al., 2017). Recently, a cocktail of hydrolytic and oxidizing enzymes from fungal consortium has been proposed for simultaneous pretreatment and saccharification, with laccase functioning effectively as a detoxifying agent (Dhiman et al., 2015). In order to degrade insoluble keratins, specific proteolytic enzymes are involved. These are termed keratinases, classified as serine- or metalloproteases and isolated from certain bacteria, actinomycetes, and fungi (Brandelli et al., 2010). While degrading materials like feathers and wool, a higher operation temperature is required and thus thermostable keratinase should be used exhibiting higher reactivity (due to lower diffusional restrictions) and enhanced stability (Nigam, 2013).

Although the biological pretreatment has many advantages, such as low energy input, lack of necessary chemicals, minimal by-product formation, and mild operating conditions, this method is time- and space-demanding and requires at least 10–14 days of residence time which results in higher reactor volumes (Rodriguez et al., 2017). In addition, the enzymatic method can be affected by many factors which have to be carefully controlled. Among them, the type of substrate, system configuration, incubation time, moisture content, and operational conditions (e.g. temperature, pH) should be involved. Since the enzymes must be regularly supplied to AD system, their application is less efficient and effective than cultivation of microbial consortia capable of producing compounds which degrade lignocellulosic compounds in a stable and continuous mode comprising various bacteria and fungi (Parawira, 2011). The advantages attributable to consortia application include their enhanced adaptability and productivity as well as the improved enzymatic efficiency, and the increased substrate utilization.

The recent progress in biological pretreatments comprises all the technological trends. In terms of low-cost method, several fungal strains have been investigated with the objective to degrade different lignocellulosic agricultural and forest residues. Liu et al. (2017) used white-rot fungi Ceriporiopsis subvermispora (strains ATCC 90467 and ATCC 96608) with a high lignin-degrading selectivity to pretreat hazel branches, acacia branches, barley straw and sugarcane bagasse, and improve their subsequent anaerobic digestion. The ligninolytic enzymes such as manganese peroxidase and laccase were produced during incubation of the samples inoculated by C. subvermispora at 28°C for 28 days. However, only in the case of hazel branches 20–25% of lignin was degraded, and thus methane production per unit mass of dry solids almost doubled the value for sample before incubation. Significant lignin degradation was not observed in the remaining samples. Rouches et al. (2016) investigated the possibility of enhancing methane yield from wheat straw pretreated with selected fungal strains. Among twelve strains tested, Polyporus brumalis BRFM 985 was found to be most efficient for pretreating the substrate. Consequently, anaerobic digestion of wheat straw was improved yielding up to 21% more methane per gram of initial total solids as compared to the control sample. It is worth noting that efficient delignification required low amounts of glucose and nitrogen addition as a starter.

In the keratin degradation area, Bacillus sp. C4 (known to produce both α- and β-keratinases) was investigated for chicken feathers pretreatment (Patinvoh et al., 2016). The results indicated that about 75.5% of the feather keratin was converted to soluble crude protein after 8 days of degradation (at initial feather concentration of 5%). Both the feather hydrolysate and the total broth (liquid and solid fraction of pretreated feathers) were supplied to the anaerobic digesters as the substrates for biogas production. Interestingly, anaerobic sludge and bacteria granules were used as inoculum. Pretreatment enhanced methane yield from hydrolysate by 292 and 105% for the first and the second type of inoculum, respectively. The same trend occurred for the total broth, in this case the associated increases were 237 and 124%, respectively.

Following another trend, the researchers constructed microbial consortia dedicated for degrading different recalcitrant waste. Poszytek et al. (2016) tested a microbial consortium with high cellulolytic activity (MCHCA) for maize silage decomposition. The consortium included 16 selected strains (representatives of *Bacillus, Providencia*, and *Ochrobactrum* genera) which had a high endoglucanase activity and exhibited wide tolerance to various physical, and chemical conditions. The MCHCA was found to be capable of efficient hydrolysis of the substrate. This resulted in an increase of biogas production by up to 38%. Thermophilic microbial consortium (MC1) with great lignocellulose degradation ability was involved by Yuan et al. (2016) for enhancement of methane yield as well as methane production rate from cotton stalk. The MC1 contained Clostridium straminisolvens (CSK1), Clostridium sp. FG4b, Pseudoxanthomonas sp. strain M1-3, Brevibacilus sp. M1-5, and Bordetella sp. M1-6. The predominant volatile organic products in the MC1 hydrolysate were ethanol, acetic acid, propionic acid, and butyric acid. It was shown that biogas and methane yields were significantly increased following MC1 pretreatment. Moreover, the methane production rate was enhnaced. In contrast, Baba et al. (2016) used a natural consortium of microbes present in cattle rumen fluid to improve the methane production from rapeseed (Brassica napus L.). The cattle rumen fluid contains 1010–1011 bacteria per gram and exhibits a high capacity of converting lignocellulose to saccharides and short-chain fatty acids. Hence, using such a consortium for waste pretreatment could result in an effective substrate solubilization and, subsequently, in enhanced biogas production. It was shown that methane production from solubilized rapeseed exceeded the values obtained from untreated substrate by 1.5 times.

While analyzing microbial community during rumen fluid treatment, some changes were observed, i.e. predominant phylum shifted from Bacteroidetes, composed of amylolytic Prevotella spp., to Firmicutes, composed of cellulolytic and xylanolytic Ruminococcus spp., in only 6 h. Moreover, 7 cellulolytic, 25 cello-oligosaccharolytic, and 11 xylanolytic bacteria were detected while investigating the most abundant sequences of detected taxa.

Constructed microbial consortia were involved in keratin decomposition. Izgaryshev et al. (2015) selected microorganisms to create a consortium for hydrolyzing residues from poultry industry and converting keratin materials into protein supplement. The consortium consisted of four selected strains showing the maximum keratinase activity: Bacillus licheniformis B-2986, Streptomyces ornatus S-1220, Penicillium rubrum F-601, and Verticillium lateritum F-626. It was found that all of the strains are resistant to lysozyme. However, the inhibitory effect was not observed when they were cocultured on solid medium with ratio of 1:1:1:1, at 37°C and pH 7.0. The conditions for poultry waste destruction were as follows: temperature – 37°C, pH 7.5, and duration of cultivation amounting to 12.0 h.

Among enzymatic pretreatments, Nagula and Pandit (2016) tested laccase for Napier grass delignification. The researchers optimized the pretreatment conditions, reporting the enzyme concentration of 10 U/gm of substrate mentioned and the impeller speed of 300 rpm. In laccase using, however, the major drawback was only 50% delignification with cellulose, since the enzyme preferentially targeted lignin, leaving cellulose intact. Gegeckas et al. (2015) cloned a keratinolytic proteinase from thermophilic bacterium Geobacillus stearothermophilus AD-11 to decompose keratinous waste. Recombinant keratinolytic proteinase (RecGEOker) with molecular weight of 57 kDa was found to effectively degrade keratin at pH 9 and 60°C. The highest substrate specificity toward keratin from wool > collagen > sodium caseinate > gelatin > and bovine serum albumin (BSA) occurred in descending order.

6 COMBINED PRETREATMENT

Different combinations of individual technologies have recently been examined in order to obtain more efficient solubilization of the recalcitrant biomass or to modify the composition of the products. Generally, two types of the strategies can be involved: simultaneously carried out processes or stepwise approach.

Kaur & Phutela (2016) studied the effect of simultaneous combination of the chemical and microwave pretreatments of paddy straw. The samples of the straw were suspended in different solutions of NaOH (2–10%) and placed in a microwave oven, where they were irradiated with microwaves (720 W) for 30 min. The digestibility of the straw and yield of the biogas production were examined. The results were compared with untreated and chemical alone treated sample. It was found that the highest decrease in lignin content equal to 65.0% was obtained during the pretreatment with 4% NaOH supplemented with microwaves. These changes entailed ca. 55% increase in biogas production. Combination of chemical and thermochemical (alkaline and diluted acid) pretreatment with stem explosion were examined by Capecchi et al. (2016). Several versions of pretreatments, including steam explosion alone (at 195°C for 5, 10 and 15 min) and after impregnation with $Ca(OH)_2$ at ambient and high (205°C) temperature and after soaking in 0.2% solution of H_2SO_4 at the temperature 195°C for 10 min were used for an improvement of switchgrass digestibility. However, in this experiment the combined methods did not prove to be the most effective. The highest lignin removal (35%) was obtained after treatment with lime at the concentration of 0.7%.

Combination of biological other pretreatment with other methods has been also considered in the recent publications. Alexandropoulou et al. (2016) have examined the effect of stepwise combination of biological treatment with white rot fungi Leiotrametes menziesii or Abortiporus biennis and alkaline method with NaOH on digestibility of willow sawdust. The maximum biochemical methane potential was observed for the biomass pretreated with the combination of NaOH and A. biennis. The potential was 12.5% and 50.1% higher than the value obtained for the sample pretreated with alkaline or fungal method alone, and 115% higher than the potential of untreated material. Mustafa et al. (2017) stated that combination of physical pretreatment of rice straw by milling with biological decomposition by fungi is a very efficient method of improving the biogas yield. When the straw was milled (≤2 mm) prior to the fungal pretreatment with Pleurotus ostreatus at 30 days, the methane yield increased by 165% in comparison to the untreated rice straw. Taking into account the energy demand the association of fungal pretreatment with the others methods is effective, but the long time of the process is the crucial disadvantage (Shirkavanda et al., 2016). Further research is needed to reduce the time of the fungal-physicochemical pretreatment.

7 CONCLUDING REMARKS

Low bioavailability of recalcitrant compounds included in some types of waste is a key factor limiting their use for biofuels production and leading to a significant decrease in the efficiency of the technological systems and reducing the profitability of the process. Numerous methods of changing the properties of biomass in order to increase its solubility that have been examined so far, differ in terms of a mechanism of the process, the efficiency of lignin removal and release of cellulose, hemicellulose, reducing sugars, toxic products, as well as the complexity of technological systems and energy demand. Moreover, the efficiency of particular methods differs considerably depending on the type of biomass, which results from its varied chemical structure. Despite the tremendous experience gained in the field of pretreatment of hardly-biodegradable biomass used for material or energy recovery, it is not clear which method is the most beneficial in terms of intensifying the biogas production. The choice of the method mainly depends on the chemical structure of the biomass and must provide a positive energy balance of the process. The amount of energy consumed in the process of fuel production cannot be greater than the amount of the energy potentially recovered from it. Except for the economic criteria, the environmental impact of the process should also be taken into account.

The chemical composition of the trace end products, like furfural or phenols, and their concentration is also a very important criterion of the evaluation the suitability of the pretreatment method in biogas production. The probability of producing the compounds that may inhibit methanogenesis varies depending on the pretreatment method. Therefore, the assessment of the suitability of the method must be preceded by methane yield tests. Not all known biomass pretreatment methods have been tested in this respect. For others, the optimum treatment conditions for particular biomass types have not been established yet. Additionally, there are a number of cause-and-effect relationships in terms of an explanation of the mechanisms of observed phenomena that need to be clarified. For example, the influence of the concentration of particular by-products that are known to be toxic for living organisms, produced during the decomposition of biomass on the consortium of microorganisms in anaerobic digester, and their adaptability to new environmental conditions.

REFERENCES

Abdul, P.M., Jahim, J. Md., Harun, S., Markom, M., Lutpi, N.A., Hassan, O., Balan, V., Dale, B.E. & Nor M.T.M. 2016. Effects of changes in chemical and structural characteristic of ammonia fibre expansion (AFEX) pretreated oil palm empty fruit bunch fibre on enzymatic saccharification and fermentability for biohydrogen. *Bioresource Technol.*, 211: 200–208.

Alexandropoulou, M., Antonopoulou, G., Fragkou, E., Ntaikou, J. & Lyberatos, G. Fungal pretreatment of willow sawdust and its combination with alkaline treatment for enhancing biogas production. *J. Environ. Manage.*, Available online 11 April 2016 (in press).

Alvira, P., Tomás-Pejó, E., Ballesteros, M. & Negro, M.J. 2010. Pretreatment technologies for an efficient bioethanol production process based on enzymatic hydrolysis: a review. *Bioresource Technol.*, 101: 4851–4861.

Arrojo, S. & Benito, Y. 2008. A theoretical study of hydrodynamic cavitation, Ultrason. *Sonochem.*, 15: 203–211.

Baba, Y., Matsuki, Y., Mori, Y., Suyama, Y., Tada, C., Fukusa, Y., Saito, M. & Nakai, Y. 2016. Pretreatment of lignocellulosic biomass by cattle rumen fluid for methane production: bacterial flora and enzyme activity analysis. *J. Biosci. Bioeng.*, 123(4): 489–496.

Barone, J.R., Schmidt, W.F. & Gregoire, N.T. 2006. Extrusion of feather keratyn. *Applied Polymer Science*, 100(215): 1432–1442.

Baugh, K.D. & McCarty, P.L. 1988. Thermochemical pretreatment of lignocellulose to enhance methane fermentation: I. Monosaccharide and furfurals hydrothermal decomposition and product formation rates. *Biotechnol. Bioeng.* 31(1): 50–61.

Bhavsar, P., Zoccola, M., Patrucco, A., Montarsolo, A., Rovero, G. & Tonin, C. 2016. Comparative study on the effects of superheated water and high temperature alkaline hydrolysis on wool keratin. *Text. Res. J.*, Available online 7 Jul 2016 (in press).

Bhutto, A.W., Qureshi, K., Harijan, K., Abro, R., Abbas, T., Bazim, A.A., Karim, S. & Yu, G. 2017. Insight into progress in pre-treatment of lignocellulosic biomass. *Energy*, 122: 724–745.

Bobleter, O. 1994. Hydrothermal degradation of polymers derived from plants. *Prog. Polym. Sci.*, 19(5):797–841.

Bolado-Rodríguez, S., Toquero, C., Martín-Juárez, J., Travaini, R. & García-Encina, P.A. 2016. Effect of thermal, acid, alkaline and alkaline-peroxide pretreatments on the biochemical methane potential and kinetics of the anaerobic digestion of wheat straw and sugarcane bagasse. *Bioresource Technol.* 201: 182–190.

Brandelli, A., Daroit, D.J. & Riffel, A. 2010. Biochemical features of microbial keratinases and their production and applications. *Appl. Microbiol. Biotechnol.*, 85: 1735–1750.

Capecchi, L., Galbe, M. Wallberg, O., Mattarelli, P. & Barbanti, L. 2016. Combined ethanol and methane production from switchgrass (Panicum virgatum L.) impregnated with lime prior to steam explosion. *Biomass Bioenerg.*, 90: 22–31.

Castoldi, R., Bracht, A., de Morais, G.R., Baesso, M.L., Correa, R.C.G., Peralta, R.A., Moreira, R.F.P.M., Polizeli, M.T., de Souz, C.G.M. & Peralta, R.M. 2014. Biological pretreatment of Eucalyptus grandis sawdust with white-rot fungi: study of degradation patterns and saccharification kinetics. *Chem. Eng. J.*, 258: 240–246.

Castro, F.B., Hotten, P.M. & Ørskov, E.R. 1994. Inhibition of Rumen microbes by compounds formed in the steam treatment of wheat straw. *Bioresource Technol.*, 50(1): 25–30.

Cheng J.J. & Timilsina G.R. 2012. Status and barriers of advanced biofuel technologies: a review. *Renew. Energy*, 36(12): 3541–3549.

Daroit, D.J., Corrêa, A.P.F. & Brandelli, A. 2009. Keratinolytic potentialof a novel Baccilus sp. P45 isolated from the Amazon basin fish Piaractus mesopotamicus. *Int. Biodeter. Biodegrad.*, 63, 358–363.

Dhiman, S.S, Haw, J., Kalyani, D., Kalia, V.C., Kang, Y.C. & Lee, J. 2015. Simultaneous pretreatment and saccharification: green technology for enhanced sugar yields from biomass using a fungal consortium. *Bioresouce Technol.*, 179, 50–57.

Du, W., Yu, H., Song, L., Zhang, J., Wenig, C., Ma, F. & Zhang, X. 2011. The promising effects of by-products from Irpex lacteus on subsequent enzymatic hydrolysis of bio-pretreated corn stalks. *Biotechnol. Biofuels*, 4: 37.

Fellahi, S., Zaghloul, T.I., Feuk-Lagerstedt, E. & Taherzadeh, M.J. 2014. A bacillus strain able to hydrolyze alpha- and beta-keratin. *J. Bioprocess. Biotech.*, 4(7): 181–188.

Fox, M. & Noike, T. 2004. Wet oxidation pretreatment for the increase in anaerobic biodegradability of newspaper waste. *Bioresource Technol.*, 91(3): 273–281.

Gao, J., Chen, L., Yan, Z. & Wang, L. 2013. Effect of ionic liquid pretreatment on the composition, structure and biogas production of water hyacinth (Eichhornia crassipes). *Bioresource Technol.*, 132: 361–364.

Gegeckas, A., Gudiukaitė, R., Debski, J. & Citavicius, D. 2015. Keratinous waste decomposition and peptide production by keratinase from Geobacillus stearothermophilus AD-11. *Int. J. Biol. Macromol.*, 75, 158–165.

Ghasimi, D.S.M., Aboudi, K., de Kreuk, M., Zandvoort, M.H & van Lier, J.N. 2016. Impact of lignocellulosic-waste intermediates on hydrolysis and methanogenesis under thermophilic and mesophilic conditions. *Chem. Eng. J.*, 295: 181–191.

Hagos, K., Zong, J., Li, D. & Lu, X. Anaerobic co-digestion process fir biogas production: Progress, challenges and perspectives. *Renew. Sust. Eng. Rev.* (article in press), http://dx.doi.org/10.1016/j.rser.2016.11.184.

Hashimoto, A.G. 1986. Pretreatment of wheat straw for fermentation to methane. *Biotechnol. Bioeng.*, 28(12): 1857–1866.

Hassan Dar, G. & Tandon, S.M. 1987. Biogas production from pretreated wheat straw, lantana residue, apple and peach leaf litter with cattle dung. *Biological Wastes*, 21 (2): 75–83

Hassan, M., Ding, W., Bi, J., Mehryar, E., Talha, Z.A.A. & Huang, H. 2016. Methane enhancement through oxidative cleavage and alkali solubilization pre-treatments for corn stover with anaerobic activated sludge. *Bioresource Technol.*, 200: 405–412.

Himmel, M.E. & Picataggio, S.K. 2009. Our challenge is to acquire deeper understanding of biomass recalcitrance and conversion. *Biomass Recalcitrance*, 1–6, Blackwell Publishing Ltd.

Izgaryshev, A.V., Krieger, O.V., Milentyeva, I.S. & Mitrokhin, P.V. 2015. Collection of microorganisms to create a consortium for the biodegradation of keratin-containing raw materials from poultry industry waste. *Biol. Med.* (Aligarh) 7(2): BM-080-15, 5 pages.

Jiang, D., Ge, X., Zhang, Q. & Li, Y. 2016. Comparison of liquid hot water and alkaline pretreatments of giant reed for improved enzymatic digestibility and biogas energy production, *Bioresource Technol.*, 216: 60–68.

Jiang, J., Wang, J., Zhang, X. & Wolcott, M. 2017. Characterization of micronized wood and energy-size relationship in wood comminution. *Fuel Process. Technol.* 161, 76–84.

Kaur, K. & Phutela, U.G. 2016. Enhancement of paddy straw digestibility and biogas production by sodium hydroxide-microwave pretreatment. *Renew. Energ.*, 92: 178–184.

Khoshnevisan, B., Shafiei, M., Rajaeifar, M.A. & Tabatabaei, M. 2016. Biogas and bioethanol production from pinewood pre-treated with steam explosion and N-methylmorpholine-N-oxide (NMMO): A comparative life cycle assessment approach. *Energy*, 114(1): 935–950.

Kondusamy, D. & Kalamdhad, A.S. 2014. Pre-treatment and anaerobic digestion of food waste for high rate methane production – a review. *J. Environ. Chem. Eng.*, 2: 1821–1830.

Li, C., Liu, G., Nges, I.A. & Liu, J. 2016a. Enhanced biomethane production from Miscanthus lutarioriparius using steam explosion pretreatment. *Fuel*, 179: 267–273.

Lissens, G. Thomsen, A.B., De Baere, L., Verstraete, W. & Ahring, B.K. 2004. Thermal Wet Oxidation Improves Anaerobic Biodegradability of Raw and Digested Biowaste. *Environ. Sci. Technol.,* 38 (12): 3418–3424.

Liu, X., Hiligsmann, S., Gourdon, R. & Bayard, R. 2017. Anaerobic digestion of lignocellulosic biomasses pretreated with Ceriporiopsis subvermispora. *J. Environ. Manage.*, 193, 154–162.

Lizasoain, J., Rincón, M., Theuretzbacher, F., Enguídanos, R., Nielsen, P.J. Potthast, A., Zweckmair T, Gronauer A., Bauer A. 2016. Biogas production from reed biomass: Effect of pretreatment using different steam explosion conditions. *Biomass Bioenerg.*, 95: 84–91.

Madison, M.J., Coward-Kelly, G., Liang, C., Karim, M.N., Falls, M. & Holtzapple, M.T. 2017. Mechanical pretreatment of biomass – part I: acoustic and hydrodynamic cavitation. *Biomass Bioenerg.*, 98, 135–141.

Maurya, D.P, Singla, A. & Negi, S. 2015. An overview of key pretreatment processes for biological conversion of lignocellulosic biomass to bioethanol. *3 Biotech.*, 5(5):597–609.

Mazotto, A.M., Couri, S., Damaso, M.C.T. & Vermelho, A.B. 2013. Degradation of feather waste by Aspergillus niger keratinases: comparison of submerged and soli-state fermentation. *Int. Biodeter. Biodegr.*, 85: 189–195.

Montusiewicz, A. 2008. Environmental factors affecting the biomethanization process. *Arch. Environ. Prot.*, 34(3): 265–279.

Montusiewicz, A., Pasieczna-Patkowska, S., Lebiocka, M., Szaja, A. & Szymańska-Chargot, M. 2017. Hydrodynamic cavitation of brewery spent grain diluted by wastewater. *Chem. Eng. J.* 313, 946–956

Mupondwa, E., Li, X., Tabil, L., Sokhansanj, S. & Adapa, P. 2017. Status of Canada's lignocellulosic etanol: Part I: Pretreatment Technologies. *Renew. Sust. Energ. Rev.*, 72, 178–190

Mustafa, A.M, Poulsen, T.G., Xia, Y. & Sheng, K. 2017. Combinations of fungal and milling pretreatments for enhancing rice straw biogas production during solid-state anaerobic digestion. *Bioresource Technol.* 224: 174–182.

Mutschlechner, M., Illmer, P. & Wagner, A.O. 2015. Biological pre-treatment: Enhancing biogas production using the highly cellulolytic fungus Trichoderma viride. *Waste Manage.,* 4: 98–107.

Nagula, K.N. & Pandit, A.B. 2016. Process intensification of delignification and enzymatic hydrolysis of delignified cellulosic biomass using various process intensification techniques including cavitation. *Bioresour. Technol.*, 213, 162–168.

Nigam, P.S. 2013. Microbial enzymes with special characteristics for biotechnological applications. *Biomolecules*, 3(3), 597–611, doi:10.3390/biom3030597.
Ostovareh, S., Karimi, K. & Zamani, A. 2015. Efficient conversion of sweet sorghum stalks to biogas and ethanol using organosolv pretreatment, Industrial Crops and Products, 66: 170–177.
Parawira, W., 2011, Enzyme research and applications in biotechnological intensification of biogas production. *Crit. Rev. Biotechnol.,* 32: 172–186.
Park, J.H., Kim, S.H., Park, H.-D., Lim, D.J. & Yoon, J.-J. 2013. Feasibility of anaerobic digestion from bioethanol fermentation residue. *Bioresource Technol.*, 141: 177–183.
Patel, V., Desai, M. & Madamwar, D. 1993. Thermochemical pretreatment of water hyacinth for improved biomethanation. *Appl. Biochem. Biotech.*, 42(1): 67–74.
Patinvoh, R.J., Feuk-Lagerstedt, E., Lundin, M., Sárvári Horváth, I. & Taherzadeh, M.J. 2016. Biological pretreatment of chicken feather and biogas production from total broth. *Appl. Biochem. Biotechnol.,* 180: 1401–1415.
Patinvoh, R.J., Osadolor, O.A., Chandolias, K., Sárvári Horváth, I.S. & Taherzadeh, M.J. 2017. Innovative pretreatment strategies for biogas production. *Bioresour. Technol.*, 224: 13–24
Peces, M., Astals, S. & Mata-Alvarez, J. 2015. Effect of moisture on pretreatment efficiency for anaerobic digestion of lignocellulosic substrates. *Waste Manage.*, 46: 189–196.
Pecorini, I., Baldi, F., Carnevale, E.A. & Corti, A. 2016. Biochemical methane potential tests of different autoclaved and microwaved lignocellulosic organic fractions of municipal solid waste. *Waste Manage.*, 56: 143–150.
Pellera, F-M. & Gidarakos, E. 2017. Microwave pretreatment of lignocellulosic agroindustrial waste for methane production. *J. Environ. Chem. Eng.*, 5: 352–365.
Perez-Pimienta, J.A. Flores-Gómez, C.A., Ruiz, H.A., Sathitsuksanoh, N, Balan, V., da Costa Sousa, L., Dale, B.E, Singh, S. & Simmons, B.A. 2016. Evaluation of agave bagasse recalcitrance using AFEX™, autohydrolysis, and ionic liquid pretreatments. *Bioresource Technol.*, 211: 216–223.
Poszytek, K., Ciezkowska, M., Sklodowska, A. & Drewniak, L. 2016. Microbial consortium with high cellulolytic activity (MCHCA) for enhanced biogas production. *Front. Microbiol.*, 7: 324.
Raud, M., Olt, J & Kikas, T. 2016. N2 explosive decompression pretreatment of biomass for lignocellulosic ethanol production. *Biomass Bioenerg.*, 90: 1–6.
Rodriguez, C., Alaswad, A., Benyounis, K.Y. & Olabi, A.G. 2017. Pretreatment techniques used in biogas production from grass. *Renew. Sust. Eng. Rev.,* 68: 1193–1204.
Rouches, E., Herpoël-Gimbert, I., Steyer, J.P. & Carrere, H. 2016. Improvement of anaerobic degradation by white-rot fungi pretreatment of lignocellulosic biomass: a review. *Renew. Sust. Eng. Rev.*, 59: 179–198.
Rouches, E., Zhou, S., Steyer, J.P. & Carrere, H. 2016. White-rot fungi pretreatment of lignocellulosic biomass for anaerobic digestion: impact of glucose supplementation. Process Biochem., 51: 1784–1792.
Salminen, E., Einola, J. & Rintala, J. 2003. The methane production of poultry slaughtering residues and effects of pre-treatments on the methane production of poultry feather. *Environ Technol.,* 24(9): 1079–86.
Schmidt, A.S. & Thomsen, A.B. 1998. Optimization of wet oxidation pretreatment of wheat straw. *Bioresour Technol.*, 64(2):139–51.
Shavandi, A, Bekhit, A.E.A, Carne, A. & Bekhit, A. 2016. Evaluation of keratin extraction from wool by chemical methods for bio-polymer application. *J. Bioact. Compat. Pol.*, 32 (2): doi: 10.1177/0883911516662069.
Shirkavand, E., Baroutian, S., Gapes, D.J. & Young, B.R. 2016. Combination of fungal and physicochemical processes for lignocellulosic biomass pretreatment – A review. *Renew. Sust. Energ. Rev.,* 54: 217–234.
Siciliano, A., Stillitano, M.A. & De Rosa, S., 2016. Biogas production from wet olive mill wastes pretreated with hydrogen peroxide in alkaline conditions. *Renew Energ.*, 85: 903–916.
Sindhu, R, Binod, P, Janu, K, Sukumaran, R. & Pandey, A. 2012. Organosolvent pretreatment and enzymatic hydrolysis of rice straw for the production of bioethanol. *World J. Microbiol. Biotechnol.*, 28(2): 473–83.
Sindhu, R., Binod, P. & Pandey, A. 2016. Biological pretreatment of lignocellulosic biomass – an overwiew, Bioresour. *Technol.* 199: 76–82.
Singh, S., Simmons, B.A. & Vogel, K.P. 2009. Visualization o of biomass solubilization and cellulose regeneration during ionic liquid pretreatment of switchgrass. *Biotechnol. Bioeng.*, 104: 68–75.
Solé-Bundó, M., Eskicioglu, C., Garfí, M., Carrère, H. & Ferrer, I. Anaerobic co-digestion of microalgal biomass and wheat straw with and without thermo-alkaline pretreatment. *Bioresource Technol.*, Available online 27 March 2017 doi.org/10.1016/j.biortech.2017.03.151 (in press).

Sun, S., Sun, S., Cao, X. & Sun, R.. 2016. The role of pretreatment in improving the enzymatic hydrolysis of lignocellulosic materials. *Bioresour. Technol.,* 199: 49–58.

Tonin, C., Zoccola, M., Aluigi, A. Varesano, A. Montarsolo, A. Vineis, C. & Zimbardi, F. 2006. Study on the Conversion of Wool Keratin by Steam Explosion. *Biomacromolecules,* 7 (12): 3499–3504.

Uellendahl, H., Wang, G., Møller, H.B., Jørgensen, U., Skiadas, I.V., Gavala, H.N. & Ahring, B.K., 2008. Energy balance and cost-benefit analysis of biogas production from perennial energy crops pretreated by wet oxidation. *Water Sci. Technol.,* 58 (9): 1841–1847.

Vincent, J. *Structural Biomaterials*; Princeton University Press, New Jersey, 1990.

Wagner, A.O, Lins, P., Malin, C., Reitschuler, C. & Illmer, P. 2013. Impact of protein-, lipid- and cellulose-containing complex substrates on biogas production and microbial communities in batch experiments. *Sci. Total. Environ.*, 458–460: 256–266.

Wang, F., Zhang, D., Wu, H., Yi, W., Fu, P., Li, Y. & Li, Z. 2016a. Enhancing biogas production of corn stover by fast pyrolysis pretreatment. *Bioresource Technol.*, 218: 731–736.

Wang, K., Li, R., Ma, J.H., Jian, Y.K & Che, J.N.. 2016b. Extracting keratin from wool by using L-cysteine. *Green Chem.*, 18 (2): 476–481. doi: 10.1039/C5GC01254F.

Wyman, C.E., Dale, B.E., Elander, R.T., Holtzapple, M., Ladisch, M.R. & Lee, Y.Y. 2005. Coordinated development of leading biomass pretreatment technologies. *Bioresource Technol.*, 96: 1959–1966.

Xia, Y., Massé, D.I., McAllister, T.A., Beaulieu, C. & Ungerfeld, E. 2012. Anaerobic digestion of chicken feather with swine manure or slaughterhouse sludge for biogas production. *Waste Manage.*, 32(3): 404–409.

Yuan, X., Ma, L., Wen, B., Zhou, D., Kuang, M., Yang, W. & Cui, Z. 2016. Enhancing anaerobic digestion of cotton stalk by pretreatment with a microbial consortium (MC1). *Bioresource Technol.*, 207, 293–301.

Zhang, K., Pei, Z. & Wang, D. 2016. Organic solvent pretreatment of lignocellulosic biomass for biofuels and biochemicals: A review. *Bioresource Technol.*, 199: 21–33.

Zhao, Z., Li, N., Bhutto, A.W., Abdeltawab, A.A., Al-Deyab, S.S., Liu, G., Xiaochun, C. & Guangren, Y. 2016. N-methyl-2-pyrrolidonium-based Brönsted-Lewis acidic ionic liquids as catalysts for the hydrolysis of cellulose. *Sci. China Chem.*, 59(5): 564–70.

Zheng, Y., Zhao, J., Xu, F. & Li, Y. 2014. Pretreatment of lignocellulosic biomass for enhanced biogas production. *Prog. Energy Combust. Sci.*, 42: 35–53.

Zhong, W., Zhang, Z., Luo, Y., Sun, S., Qiao, W. & Xiao, M. 2011. Effect of biological pretreatments in enhancing corn straw biogas production. *Bioresource Technol,*. 102: 11177–11182.

Zhou, X., Li, Q., Zhang, Y. & Gu, Y. 2017. Effect of hydrothermal pretreatment on Miscanthus anaerobic digestion. *Bioresource Technol.*, 224: 721–726.

Zuroff, T.R., Xiques, S.B. & Curtis, W.R. 2013. Consortia-mediated bioprocessing of cellulose to ethanol with a symbiotic Clostridium phytofermentans/yeast co-culture. *Biotechnol. Biofuels*, 6: 59, doi: 10.1186/1754-6834-6-59.

Problems with energy supply

Z. Cao
Institute of Soil Science, Nanjing, China

L. Pawłowski & A. Duda
Lublin University of Technology, Lublin, Poland

ABSTRACT: The role of energy acquisition in the implementation of sustainable development idea is discussed. The impact of biofuels production on the degradation of the environment, with CO_2 emission taken into account, was discussed in detail. It was shown, that the production of biofuels from agricultural crops causes an increase in food prices. The carbon cycle in our biosphere was described as well, showing that the natural fluxes are much more important than anthropological ones. The potential of Poland in bio-sequestration of carbon dioxide was also analyzed.

Keywords: sustainable development; energy supply; climate change; greenhouse gases

1 INTRODUCTION

The supply of primary energy and processing it into usable energy forms exerts a huge impact on the economic, environmental, and social aspect of civilizational development. The technological revolution, which initiated a rapid and on-going development of the world, has mainly been based on the use of fossil fuels. The problem is that the issue of energy supply is most frequently considered in the context of CO_2 emissions. Some people advocate that the use of fossil fuels, especially coal, should be ceased entirely. Meanwhile, according to the "Energy International Agency" programme, coal will continue to be the basic energy resource for another several dozen years and its deposits should suffice for about 150 years (Figure 1) (Konstańczak 2014). The estimated reserves of crude oil and gas, with the current rate of consumption, will be enough for approximately 50–60 years.

According to the forecast energy mix, in 2050 coal will constitute as much as 43% of the consumed primary energy resources (Table 1) (La Quere et al. 2015). Taking these forecasts into the account, focusing on mitigation of greenhouse gases emissions by resigning from

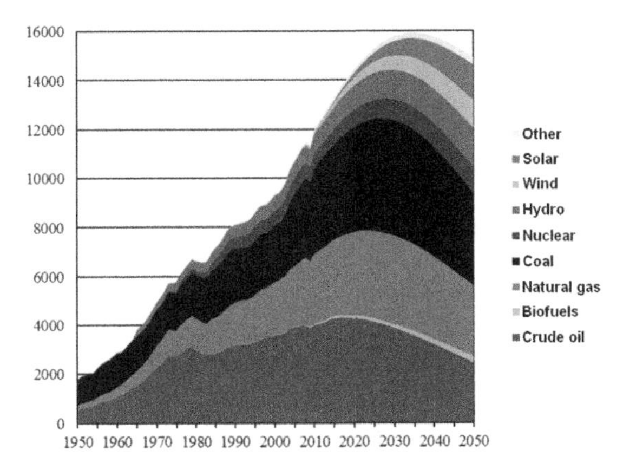

Figure 1. Consumption of primal energy resources in the years 1950–2015 and the forecast consumption for 2015–2050 (La Quere et al. 2015).

Table 1. Forecast consumption of primary energy resources in 2015.

Coal	43%
Wind	12%
Hydro	12%
Petrol	9%
Gas	9%

coal as the main energy resource is not feasible. However, this does not mean that alternative energy sources should not be developed. New solutions should be examined carefully, utilizing a full Life Cycle Analysis of a prospective energy source (Sztumski 2016, Cel et al. 2016, Macuda et al. 2017, Velkin et al. 2017).

2 PROBLEMS CONNECTED WITH BIOFUELS

In line with resolution of the EU Commission made in 2009, as much as 10% of energy used in transportation should mainly come from the biofuels obtained from agricultural crops. In order to make biofuels profitable, governments subsidize their production. In the USA, as much as 127 million tons of corn, i.e. 40% of annual production was allocated for bioethanol production in 2011. Production of ethanol for fuel purposes was awarded 6 billion dollars in subsidies. Allocating such great amounts of corn for ethanol production resulted in a two-fold increase of corn prices over the period 2007–2012. Moreover, huge import of agricultural crops for biofuel production by the EU member countries resulted in a dramatic, 2.5-fold increase of FAO food price index in the years 2007–2010. It should be emphasized that the growth of food prices is felt especially severely by the poor.

In view of the above, utilizing biomass from the agricultural crops as a source of energy threatens the realization of sustainable development, because it violates the intragenerational equity paradigm by limiting the availability of food to the poor (Liu 2015, Konstańczak 2014, Żukowska et al. 2016, Velkin et al. 2017).

Moreover, attention should be drawn to the fact that the biofuels are promoted on the basis of an erroneous assumption that the plants only release the same amount of CO_2 during combustion, as was absorbed throughout their growth. This simplified way of thinking does not account for the entire production cycle of biofuels. These estimations do not consider the changes in land use, as well as the energy outlays required for the cultivation and processing of biomass into biofuels.

Strong pressures on the usage of biofuels in transportation resulting from the European Union policy caused that the tropical forests, especially in the developing countries, are cut down and biofuel plants are being cultivated in their place. The studies conducted by Danielsen et al. (2009) indicate that the capacity of the tropical forests to absorb CO_2 is much greater than the one of the plants used for biofuel production. Consequently, cultivation of plants for biofuels lowers the CO_2 absorption in the areas of cut tropical forests. Transforming the tropical forests and peatlands into biofuel crops leads to an additional emission of CO_2, reaching the amount of approximately 55 Mg CO_2 from 1 ha over the period of 120 years. Hence, the application of biofuels derived from agricultural crops does not mitigate the CO_2 emissions (Melillo et al. 2009).

Moreover, in order to produce a biofuel, e.g. corn bioethanol, it is necessary to spend energy on the cultivation, production of fertilizers, plant harvesting, as well as fuel processing with fermentation and distillation. Using the Life Cycle Analysis method it was shown that the amount of CO_2 emitted per unit of energy derived from corn bioethanol is 60% greater in relation to the amount of CO_2 emitted from the combustion of equivalent amount of crude oil fuels. Even Brazil, where the production of bioethanol from sugar cane is most advanced and fully utilizes the residual biomass, e.g. stalks for the production of thermal energy, failed to lower the emission of CO_2 per unit of energy below the level characterizing the liquid biofuels derived from crude oil.

Biofuels are also ambiguous in respect to their energy efficiency. On the one hand, according to the research carried out by Piementel (Pimentel 2012), the amount of energy used for the production of bioethanol is higher than the energy gained from the combustion of ethanol in automobile engines. The process of corn ethanol production consumes 29% more energy than is gained from the combustion of ethanol derived from grass – by 45% – and from wood – by 57%.

Similar situation occurs with soy biodiesel production, which consumes 27% more energy than is gained from the produced biodiesel, whereas in the case of the biodiesel produced from sunflower seeds, this value is increased to 118%.

The above-mentioned data shows that in the United States (and probably other developed countries as well) the cultivation of plants for liquid fuel production is not sustainable, because it increases the consumption of fossil fuels and CO_2 emission even further.

The negative impact of expanding areas of land allocated for the cultivation of energy crops exerted on the habitats and bio-diversity is also extremely important. It should be noted that more than the half of terrestrial animal species live in the tropical forests. The forests in south-eastern Asia, which are highly abundant in habitats of various organisms, are most threatened by the creation of biofuel plantations. Tropical forests also absorb approximately 46% of carbon dioxide contained in the atmosphere. Their destruction may lead to a 25% increase of carbon dioxide content in the atmosphere.

Therefore, there is an internal contradiction related to the allocation of the areas occupied by tropical forests for the cultivation of biofuel plants with low emission of carbon dioxide. It is estimated that converting forests into biofuel plantations would reduce the number of species inhabiting these areas by the factor of five.

Production of liquid biofuels for transportation also exerts a negative influence on the aquatic environment due to a heavy consumption of water used both for the irrigation of crops and in the processing of plants into biofuels. Notably, processing of plants yields substantial amounts of wastewater characterized by high environmental impact, for instance, producing 1 litre of ethanol simultaneously creates 6–12 litres of highly polluted wastewater. Meanwhile, the shortage of water already negatively impacts the food production.

Generally, producing 1 litre of bioethanol consumes approximately 2500 litres of water, which is equivalent amount to the one required to produce food for one person. In order to irrigate 30 000 000 hectares used for the cultivation of biofuel crops, 180 km^3 of fresh water will be needed.

It should be noted that due to the growth of population to 8.3 billion in 2030 (7.2 billion in 2012), the demand for food, water, and energy will continue to increase by 35%, 40%, and 50%, respectively.

Vast areas of monocultures, usually employed in the case of biofuel crops, require the application of substantial amounts of herbicides and pesticides, which subsequently infiltrate into the groundwater, polluting it. The soy crops in Brazil are an example of a negative impact of pesticide usage. Wide application of pesticides and herbicides threatens the wetlands of Pantanal, an important region which provides habitats for hundreds of bird, mammal, and reptile species. Another example includes sugar cane plantation (20 000 ha) allocated for ethanol production, located in the Tana River Delta in Kenya. The planned water consumption of 1680 m^3 of water/min constitutes about 30% of river flow, seriously threatening the local ecosystem, inhabited by 345 species of water and marsh birds.

The discussion shows that although biofuels undoubtedly constitute a renewable source of energy, their application is not neutral to the environment. This means that biofuels do not meet the criteria of the sustainable development.

3 THE ROLE OF CARBON

The energy supply is and will no doubt continue to be one of the most important factors governing the development or even survival of the human civilization (Yohe 2007, Banuri and Opschoor 2007, Munasinghe 2001, Konarski 2014, Beer et al. 2010).

A critical analysis of the current energy supply trends indicates that the utilizing all available sources of energy is necessary. Optimizing the technology of coal usage, leading to minimization of environmental impact, constitutes a rational supplement to the issues discussed at COP, especially from the Polish perspective (over the next few years, limiting the use of coal as a primary energy source is not possible).

One of the key issues connected with processing the primary energy resources into various forms of usable energy are the climate changes caused by excessive emission of greenhouse gases, especially the carbon dioxide. These changes threaten the realization of two most important sustainable development paradigms: intra- and intergenerational equity (Piemental 2012).

The reports of Intergovernmental Panel on Climate Change (IPCC) predict that if the emissions of CO_2 to the atmosphere are not stopped, the consequences for the climate will be dire. One of the most severe ones will involve disruption of rainfall patterns in particular climatic zones, which will negatively impact the food production. Although it is commonly assumed that the effects of climatic changes will be disastrous, one should consider the works of American climatologist Richard Lindzen (Lindzen 2010), bearing in mind that the cost related to the implementation of low-carbon technology will be enormous. Lindzen questions the scale of climate changes predicted by IPCC. This belief is important, because the technologies of CO_2 emission mitigation are connected with additional power consumption, leading to even faster exhaustion of already limited fossil fuel reserves. This, in turn, violates another paradigm of sustainable development: the intergenerational equity paradigm.

On the other hand, it should be noted that the increase of CO_2 concentration in the atmosphere can also be beneficial because it contributes to a faster growth of biomass, including food, as its rate of growth is dependent on the assimilation of CO_2 from the atmosphere.

However, it seems that the emission of CO_2 from the anthropogenic sources constitutes a small fraction of the natural CO_2 fluxes. In order to better understand the role of carbon dioxide in the environment, the carbon cycle in terrestrial ecosystems will be analyzed.

In the course of photosynthesis, approximately 455 billion tons of CO_2 is transformed into biomass annually (Le Quere 2015, IPCC 2013, Statistical Year Book 2014). Simultaneously, about 440 billion tons of CO_2 are emitted during respiration of organisms and decomposition of biomass each year. This means that the net absorption of the terrestrial ecosystems amounts to 15 billion tons of CO_2. Huge CO_2 fluxes also occur between the atmosphere and oceans and seas. The sea and ocean waters absorb 269 billion tons of CO_2, simultaneously releasing 262 billion tons of CO_2. Therefore, the net CO_2 absorption by the sea and ocean waters approximates 7 billion tons of CO_2. Unfortunately, due to acidification, the absorption of CO_2 by the sea and ocean waters will decrease.

Moreover, small amounts of CO_2 are absorbed by weathering minerals and emitted by volcanoes.

In comparison to the above-mentioned natural fluxes, the emission from the anthropogenic sources, i.e. combustion processes and cement production, is relatively low.

According to the data, the emission of CO_2 in 2014 ranged from 34.8 to 39.3 billion tons of CO_2, whereas the forecast emission in 2020 will be in the range 39.7–45.6 billion tons of CO_2 (La Quere 2015).

The second source of CO_2 emission, connected with the anthropogenic activity, is the emission caused by changes in land use (deforestation, fires, draining of swamps, etc.).

The balance of emission and absorption of CO_2 indicates that over the period 2011–2014, the emission of CO_2 to the atmosphere increased by 3.6% and thus the growing CO_2 concentration in the atmosphere is observed. For instance, in the years 1870–2014, the concentration of CO_2 in the atmosphere increased from 288 ppm to 397 ppm. This growth resulted from the combined emissions from the combustion of: coal + 89 ppm, petroleum products + 67 ppm, gas + 28 ppm and cement products + 5 ppm. On the other hand, the two main natural processes responsible for lowering the concentration of CO_2 in the atmosphere, i.e. net absorption by the terrestrial ecosystems and the net absorption by seas and oceans correspond to a decrease by 6 and 79 ppm of CO_2, respectively. A slight drop was observed in 2014. In 2015, the CO_2 emission was lower by 0.6%.

According to the above-mentioned data, the natural processes of CO_2 emission and absorption by particular terrestrial ecosystems are the dominant ones. Their role will be demonstrated on the example of Poland.

4 ASSESSMENT OF CO_2 SEQUESTRATION POSSIBILITY BY TERRESTRIAL ECOSYSTEMS IN POLAND

Emission of CO_2 from fossil fuel combustion and cement production in Poland systematically drops since 1990, mainly as a result of closure of industry due to so-called Balcerowicz Plan. In 2014, the total CO_2 emission in Poland reached 316.8 million tons (Statistical Year Book 2014).

From the point of view of mitigating CO_2 emission, the Polish forests, with an approximate area of 9.4 million ha, constitute an important ecosystem (Statistical Year Book 2014). Depending on the type and age of trees, 1 ha of forest absorbs 30–35 tons of CO_2/year. Hence, the total annual CO_2 absorption by the Polish forests ranges from 283–329 million tons of CO_2. Simultaneously, forests emit CO_2 in the course of respiration and decomposition of organic matter. Gaj (2012), drawing on the studies conducted by Veroustraele and Sabie, reports that the net CO_2 absorption by the Polish forests amounts to 9 t/ha per year on average. Similar absorption intensity – 3.1 million tons of CO_2/year – is exhibited by orchards, which occupy 341.8 thousand ha.

Significant amount of CO_2 is also absorbed by pastures (4.8 tons of CO_2/ha per year) and meadows (2.6 tons of CO_2/ha per year) (Gaj 2012, Sauerbeck 2001, Acharya et al. 2012, Mota 2010). Therefore, the annual CO_2 sequestration of CO_2 by pastures equals 1.9 million tons of CO_2/year. On the other hand, meadows extend over a much greater area, i.e. 2.6 million ha; thus, the annual CO_2 sequestration by meadows amounts to 6.8 million tons of CO_2/year.

In Poland, cereals occupy the greatest area, i.e. 7.48 million ha. The available data on CO_2 sequestration was determined for Spain (Mota 2010). They amount to 13.9 and 11.7 million tons of CO_2/year for wheat and oat, respectively. Sequestration in the Polish climatic zone will be lower. Taking this data into account, it can be estimated that the CO_2 sequestration by cereals amounts to 74.8 million tons of CO_2/year. The total CO_2 sequestration by orchards, cereals, pastures, and meadows adds up to 86.6 million tons of carbon/year, i.e. 27%. Moreover, industrial crops and potatoes cultivated in Poland occupy 1.16 million ha and 267 thousand ha, respectively. Unfortunately, no data on the CO_2 sequestration by these plants is available. It can be assumed that it will be similar to the sequestration of such vegetables as cauliflower, broccoli, artichokes, and tomatoes. Following values of CO_2 sequestration are given for these vegetables in the literature (Mota 2010): cauliflower – 36 tons of CO_2/ha per year, broccoli – 22 tons of CO_2/ha per year, artichoke – 13 tons of CO_2/ha per year, and tomato – 24 tons of CO_2/ha per year. In Poland, potatoes are cultivated on 267 thousand ha. Assuming the level of CO_2 sequestration is similar to artichokes, potatoes will absorb 3.5 million tons of CO_2/year. In the case of industrial crops, we assume the sequestration value of crops, i.e. 11 tons of CO_2/ha per year, which with 1.16 million ha would translate to 12.8 million tons of CO_2/year.

In line with the above-mentioned calculations, the CO_2 sequestration by agricultural crops equals 103 million tons of CO_2/ha per year, and 187.5 tons of CO_2/ha per year when combined with forests, which roughly constitutes 59% of emission from fuel combustion and cement production. This means that 59% of CO_2 emission from fossil fuel combustion and cement production is absorbed by agricultural crops and forests.

Moreover, CO_2 sequestration can potentially be improved by the terrestrial ecosystems. Information can be found in the literature (Wójcik et al. 2014) that in Poland there are 2.3 million ha of marginal soils. Their forestation would enable to increase the sequestration of CO_2 by 20.7 million tons of CO_2/year per year. Moreover, 175 thousand ha of land is fallow. It could be used for sequestration of CO_2 by cultivating green biomass, which would yield double benefits. It would raise the sequestration of CO_2 by additional 6.2 million tons of CO_2/year and improve the fertility of soil owing to an increase in humus content. Undoubtedly, it would reduce the need for mineral fertilizers, especially if plants that absorb nitrogen from air were used for the cultivation of green biomass on the fallow land. As a result, mitigation of greenhouse gases emission connected with the production of mineral fertilizers would also be possible.

Sequestration of CO_2 by terrestrial ecosystems constitutes an important method of mitigating the emission of CO_2 into the atmosphere. The authors are not experts on the cultivation of plants; hence, the presented data should be treated as rough estimations. Conducting a thorough analysis of the existing and prospective possibilities of CO_2 sequestration by the terrestrial ecosystems in Poland would be appropriate. This is all the more important because, as shown in the paper, the improvement of CO_2 sequestration is possible through further forestation and use of fallow lands for the cultivation of green biomass, enabling the absorption of at least 27 million tons of CO_2/year, which corresponds to as much as 8.5% of annual CO_2 emission from the combustion of fossil fuels and cement production.

5 THE ROLE OF METHANE EMISSION

While considering the contribution of CO_2 in the greenhouse effect, which amounts to 65%, one should also remember about the second important greenhouse gas, i.e. methane. Its share in the greenhouse effect equals 16% (NASA homepage). The global emission of methane from all sources combined reaches 771 million t/year. The methane emitted from the natural sources amounts to 223.6 million t/year (169.6 million t/year is emitted from swampy areas, 30.8 million t/year is produced by termites, whereas 23.1 million t/year is emitted from hydrates contained in oceans). The emission from the anthropogenic sources is much greater and reaches 547 million t/year (gas, crude oil, and coal mines produce 103.9 million t/year, husbandry of ruminants 87.5 million t/year, rice cultivation 65.6 million t/year, biomass combustion 43.8 million t/year, landfills 32.8 million t/year, wastewater and animal waste treatment plants 27.4 million t/year each).

One of the methane emission sources, directly connected with environmental engineering, is the emission from landfills, constituting 6% of the total emission from the anthropogenic sources. At landfills, methane is generated in the course of methane fermentation of biodegradable wastes (Montusiewicz et al. 2008, Staszewska & Pawłowska 2011).

Implementation of sustainable development requires utilizing the emitted methane for energy purposes to the greatest possible extent (Hoglund-Isaksson 2012). Low emission of methane can be increased by introducing sewage sludge to landfills (Pawłowska and Siepak 2006) or by employing the natural methane oxidation processes in an appropriately shaped cover (Stępniewski & Pawłowska 1996, 2006) or a passive biofilter (Pawłowska 2008).

REFERENCES

Acharya, B.S., Rasmussen, J. & Eriksen, J. 2012. Grassland Carbon Sequestration and Emissions Following Cultivation in a Mixed Crop Rotation. *Agriculture, Ecosystems and Environment*, vol. 153, 33–39.

Banuri, T. & Opschoor, H. 2007. Climate Change and Sustainable Development, DESA working Paper No 56/2007, St/ESA/2007/DWP/56.

Beer, C. et al. 2010. Terrestrial Gross Carbon Dioxide Uptake: Global Distribution and Covatiation with Climate, in: *Science*, vol. 329/2010, issue 5993, s. 834–838. doi: 10.1126/science.1184984.

Beg, N. et al. 2002. Linkages between climate change and sustainable development, in: Climate Policy 2/2002, p. 129–144.

Cel, W., Czechowska-Kosacka, A. & Zhang, T. 2016. Sustainable Mitigation of Greenhouse Gases Emissions, in: *Problemy Ekorozwoju/Problems of Sustainable Development*, 11(1): 173–176.

Chakraborty S., Sadhu P.K. & Goswami U. 2016. Barriers in the Advancement of Solar Energy in Developing countries like India. *Problemy Ekorozwoju/Problems of Sustainable Development*, 11(2): 75–80.

Chefuruka, P. 2007. Report: World Energy to 2050, Forty Years of Decline.

Danielsen, F. et al. 2009. Biofuel plantations on forested lands: double jeopardy for biodiversity and climate. *Conserv Biol.*, 23(2): 348–58. doi: 10.1111/j. 1523–1739.2008.01096.

Gaj, K. 2012. Carbon Dioxide Sequestration by Polish Forest Ecosystems. *Leśne Prace Badawcze (Forest Research Papers)*, 73(1): 17–21. doi: 10.2478/v10111-012-0002-8.

Höglund-Isaksson, L. (2012). Global anthropogenic methane emissions 2005–2030: technical mitigation potentials and costs, Atmos. Chem. Phys., 12, 9079–9096.

IPCC, 2013. Climate Change 2013: The Physical Science Basis, The Fifth Assessmetn Report of the Intergovernmental Panel on climate Change. Cambridge University Press, Cambridge. doi: 10.1017/CBO9781107415324.

Jarosz, S. 1964. *Istota i znaczenie ochrony przyrody*. Warszawa: Liga Ochrony Przyrody.

Konarski, W. 2014. Mineral energy sources and political activities: introduction to mutual dependencies and their selected exemplification. *Problemy Ekorozwoju/Problems of Sustainable Development*, 9(1): 63–70.

Konstańczak, S. 2014. Theory of sustainable development and social practice. *Problemy Ekorozwoju/Problems of Sustainable Development*, 9(1): 37–46.

Le Quere, C. et al. 2015. Global Carbon Budget 2014, im: *Earth Syst. Sci. Data*, 7: 47–85.

Lindzen, R.S. 2010. Global warming: the origin and nature of the alleged scientific consensus. *Problemy Ekorozwoju/Problems of Sustainable Development*, 5(2): 13–28.

Liu, H. 2015. Biofuel's Sustainable Development under the Trilemma of Energy, Environment and Economy. *Problemy Ekorozwoju/Problems of Sustainable Development*, 10(1): 55–59.

Macuda, J., Bogacki, M. & Siemek, J. 2017. Effect of Drilling for Shale Gas in the Quality of Atmospheric Air. *Problemy Ekorozwoju/ Problems of Sustainable Development,* 12(1): 91–100.

Meadows, D.H.,. Meadows, D.L. & Behrens, W.W. 1972. *The Limits to Growth*. New York: Universe Books.

Montusiewicz, A., Lebiocka, M. & Pawłowska, M. 2008. Characterization of the biomethanization proces in selected waste mixtures. *Archives of Environmental Protection*, 34 (3), 49–61.

Mota, C. et al. 2010. Absorption of CO_2 by the Most Representative in the Region of Murcia Crops, CSIC report.

Munasinghe, M. 2001. Exploring the Linkages between climate Change and sustainable Development: A Challenge for Transdisciplinary research. *Ecology and Society*, 5(1), art. 14.

NASA homepage. www.nasa.gov.

Pawłowska, M. & Siepak, J. 2006. Enhancement of methanogenesis at a municipal landfill site by addition of sewage sludge. *Environmental Engineering Science*, 23(4): 673–679.

Pawłowska, M. & Stępniewski, W. 2006. Biochemical reduction of methane emission from landfills. *Environmental Engineering Science*, 23(4): 666–672.

Pawłowska, M. 2008. Reduction of methane emission from landfills by its microbial oxidation in filter bed, Management of Pollutant Emission from Landfills and Sludge, Book Series: Proceedings and Monographs in Engineering Water and Earth Sciences, pages 3–20.

Piementel, D. 2012. Energy Production from Maize. *Problemy Ekorozwoju/Problems of Sustainable Development*,7(2): 15–22.

Sauerbeck, D.R. 2001. CO_2 Emissions and C Sequestration by Agriculture – Perspectives and Limitations. *Nutrient Cycling in Agroecosystems*, 60(1).

Staszewska, E. & Pawłowska, M. 2011. Characteristics of emissions from municipal waste landfills. Environment Protection Engineering, 37 (4), 119–130.

Statistical Year Book 2014.

Stępniewski, W., Pawłowska, M. et al. 1996. A possibility to reduce methane emission from landfills by its oxidation in the soil cover, Chemistry for the Protection of the Environment 2, Book Series: Environmental Science research, 51, 75–92.

Sztumski, W. 2016. The Impact of Sustainable Development on the Homeostasis of the Social Environment and the Matter of Survival. *Problemy Ekorozwoju/Problems of Sustainable Development*, 11(1): 41–47.

Tobera, P. 1988. *Kryzys środowiska, kryzys społeczeństwa, Kryzys środowiska – kryzys społeczeństwa.* Warszawa: Ludowa Spółdzielnia Wydawnicza.

Velkin V.I. & Shcheklein S.E. 2017. Influence of RES Integrated Systems on Energy Supply Improvement and Risks. *Problemy Ekorozwoju/Problems of Sustainable Development*, 12(1): 123–129.

Wójcik, J., Balawejder, M. & Leń, P. 2014. Grunty marginalne, propozycje sposobów ich zagospodarowania w pracach scaleniowych w powiecie Brzozowski. *Infrastruktura i Ekologia Terenów Wiejskich*, 2: 399–410.

Yohe, G.W. et al. 2007. Perspectives on climate change and sustainability. Climate Change 2007: Impacts, Adaptation and Vulnerability. Contribution of Working Group II to the fourth Assessment Report of the Intergovernmental Panel on Climate Change., s. 811–841.

Zdeb, M. & Pawłowska, M. 2009. An influence of temperature on microbial removal of hydrogen sulphide from biogas. *Rocznik Ochrona Środowiska*, 11: 1235–1243.

Żelazna, A. & Gołębiowska, J. 2015. Measures of Sustainable Development – a Study Based on the European Monitoring of Energy-related Indicators. *Problemy Ekorozwoju/Problems of Sustainable Development*, 10: 169–177.

Żukowska G., Myszura M., Baran S., Wesołowska S., Pawłowska M. & Dobrowolski Ł. 2016. Rolnictwa a łagodzenie zmian klimatu. *Problemy Ekorozwoju/Problems of Sustainable Development*, 11(2): 67–74.

Author index

For Product Safety Concerns and Information please contact our EU representative GPSR@taylorandfrancis.com Taylor & Francis Verlag GmbH, Kaufingerstraße 24, 80331 München, Germany

Printed and bound by CPI Group (UK) Ltd, Croydon, CR0 4YY
01/05/2025
01858565-0002